STATISTICAL METHODS IN SONAR

STUDIES IN SOVIET SCIENCE
PHYSICAL SCIENCES

1974

THEORY OF PLASMA INSTABILITIES
Volume 1: Instabilities of a Homogeneous Plasma
A. B. Mikhailovskii

THEORY OF PLASMA INSTABILITIES
Volume 2: Instabilities of an Inhomogeneous Plasma
A. B. Mikhailovskii

NONEQUILIBRIUM STATISTICAL THERMODYNAMICS
D. N. Zubarev

REFRACTORY CARBIDES
G. V. Samsonov

WAVES AND SATELLITES IN THE NEAR-EARTH PLASMA
Ya. L. Al'pert

1975

ENVIRONMENTAL HAZARDS OF METALS:
Toxicity of Powdered Metals and Metal Compounds
I. T. Brakhnova

DOMAIN ELECTRICAL INSTABILITIES IN SEMICONDUCTORS
V. L. Bonch-Bruevich, I. P. Zvyagin, and A. G. Mironov

1976

THE ROTATING DISC ELECTRODE
Yu. V. Pleskov and V. Yu. Filinovskii

1977

THEORY OF TURBULENT PLASMA
V. N. Tsytovich

THEORETICAL FOUNDATIONS OF NONLINEAR ACOUSTICS
O. V. Rudenko and S. I. Soluyan

LASER-INDUCED DISCHARGE PHENOMENA
Yu. P. Raizer

LIQUID-PHASE OXIDATION OF OXYGEN-CONTAINING COMPOUNDS
E. T. Denisov, N. I. Mitskevich, and V. E. Agabekov

1978

ADHESION OF SOLIDS
B. V. Deryagin, N. A. Krotova, and V. P. Smilga

STATISTICAL METHODS IN SONAR
V. V. Ol'shevskii

THE SILOXANE BOND
M. G. Voronkov, V. P. Mileshkevich, and Yu. A. Yuzhelevskii

STUDIES IN SOVIET SCIENCE

STATISTICAL METHODS IN SONAR

V. V. Ol'shevskii
Acoustics Institute
Academy of Sciences of the USSR
Moscow, USSR

Technical Editor
David Middleton
University of Rhode Island

CONSULTANTS BUREAU • NEW YORK AND LONDON

Library of Congress Cataloging in Publication Data

Ol'shevskii, Viktor Vladimirovich.
Statistical methods in sonar.

(Studies in Soviet science: Physical sciences)
Translation of Statisticheskie metody v gidrolokatsii.
Bibliography: p.
Includes index.
1. Sonar. I. Title. II. Series
VM480.O4713 623.89'38 78-18196
ISBN 0-306-10947-6

The original Russian text, published by Sudostroenie in Leningrad in 1973, has been corrected by the author for the present edition. This translation is published under an agreement with the Copyright Agency of the USSR (VAAP).

СТАТИСТИЧЕСКИЕ МЕТОДЫ В ГИДРОЛОКАЦИИ
В. В. Ольшевский

STATISTICHESKIE METODY V GIDROLOKATSII
V. V. Ol'shevskii

A Division of Plenum Publishing Corporation
227 West 17th Street, New York, N.Y. 10011

Printed in the United States of America

FOREWORD

Dr. V. V. Ol'shevskii is perhaps most familiar to Western readers as the author of "Characteristics of Sea Reverberation," published in translation by Consultants Bureau (New York, 1967). The present book, "Statistical Methods in Sonar," is, in part, a sequel to the first book, where now the author's stated purpose is "to acquaint a broad range of specialists with the use of contemporary statistical methods for solving theoretical and applied sonar problems." As the author quite properly observes, the work is illustrative, devoted to a variety of relevant, specific technical problems from an analytical point of view, and is not in any way intended to be an all-inclusive treatise. Nevertheless, as the reader can verify subsequently, the author has succeeded in accomplishing his stated purpose. He has, moreover, provided us with a useful and, in a number of instances, provocative work, which even five years after its original appearance retains its freshness and interest with material not to date covered in other books on the subject (for example, see Horton [40], Stephens [41]).*

In this Foreword we first concisely review the author's material, on a chapter-by-chapter basis, after which a short general critique is given. Attention is called to various topics of particular interest to the professional audience, as well as to a number of highlights which deserve the reader's notice (a few additional comments on the technical editing are then included).

First of all, the treatment is statistical and analytical. The principal specific general assumptions made throughout are:

---

*Numbers in brackets, here and in the text, refer to references in the Bibliography at the end of the book. Numbers beyond [36] have been added by the editor. (D.M.)

(i) Gaussian (i.e., high-density Poisson processes) for scattering in the medium

(ii) independent, discrete scatterers (¶36)

(iii) zero gradient oceans ($\nabla c = 0$)

(iv) omnidirectional beam patterns (¶36)

(v) narrow-band signals (mostly)

(vi) essentially monostatic operation (¶44)

The various pertinent statistics of received signal and reverberation are consequently: means, variances, and other second-order moments, such as correlation functions and spectra, along with the associated Gaussian and Rayleigh probability (density) distributions. Within these constraints a considerable body of results is then generated, along with a clear exposition of the approach, which, as expected, involves a direct application of now standard statistical communication-theoretic methods (see, for example, [12, 45-47]).

The nature of the material discussed may now be briefly described:

Chapter I: "Introductory Remarks": Here the author defines "sonar" for the purposes of this book as "active sonar," the "objects of research" as the "objects of detection," and the general aims of sonar usage here as involving detection, estimation, and classification of the objects of research. A sonar model is then specified in a general way. The main differences between sonar and radar are noted, with the following features unique to sonar: (i) much greater variability of the physical medium; (ii) more complex shapes (vis-à-vis the returned signal) of the objects of detection; (iii) presence of reverberation in the medium and from its boundaries; (iv) $c_{\text{acoustic}} \ll c_{\text{EM}}$: the much slower speed of acoustic propagation in the medium, and the lower frequencies required because of absorption, lead to much smaller data rates and much greater restrictions on information flow in the channel. The author then states that the central tasks of analyzing and synthesizing sonar systems are to develop suitable "indices of effectiveness," e.g., appropriate probabilistic measures of performance.

Chapter II: "Procedural Fundamentals of Experimental Research in Sonar": In this chapter the author presents a hierarchical ordering of qualitative and quantitative procedures for "zeroing-in" on the *base model*, i.e., the appropriate model of the particular sonar situation at hand, which is to be studied experimentally and theoretically.

Chapter III: "Probability Models of Hydroacoustic Signals": Two major approaches to signal propagation and scattering are distinguished here. These are: (i), the so-called "wave models," which involve solving (a set of) appropriate wave equations (Langevin equations), with boundary and initial conditions and ensemble properties, the goal of which is to determine the probability distributions of the random variables in question; and (ii), "phenomenological models," which employ simplified signal structures and embed the detailed boundary conditions and ensemble properties in a set of appropriately introduced random parameters or processes, to yield scattered and received waveforms with characteristic amplitude and phase (or frequency) modulations, related phenomenologically to the various random scattering mechanisms and propagation conditions involved.

While method (i) is acknowledged to be conceptually more precise and physical, it has the chief disadvantage of often exclusive analytical complexity vis-à-vis the phenomenological approach, (ii), which is the approach adopted by the author in this book. Accordingly, the focus here is on "the statistical properties of measurable signals," and a further distinction is made between "canonical," "constructive," and "parametric" phenomenological models.

Chapter IV: "The Emitted Signals": This chapter is essentially an exercise in calculating ambiguity functions for various classes of emitted narrow-band signals (under the previously mentioned assumption of omnidirectional beams). While the results are not generally new, they provide a convenient set for reference (and later use in the book). Included here are ambiguity functions for rectangular cw pulses, frequency- and phase-modulated cw pulses (linear chirp, with Gaussian envelopes, etc.), and perhaps of chief interest, signals in the form of noise pulses, for which are calculated the means and variance of the associated ambiguity functions.

Chapter V: "Echo Signals": The "echo signals" here are the desired signals reflected from the object of detection (or "research," cf. Chapter I) and received by the omnidirectional receiving array. The emphasis is on Gaussian signal variations imposed on the original, injected signal, and on the calculation of cross-ambiguity functions between these emitted and received (desired) signals. Because of the Gaussian assumptions, a second-moment analysis, involving the calculation of the mean, variance, and correlation function of the cross-ambiguity function, is developed.

Chapter VI: "Marine Reverberation": It is at this stage that the aforementioned assumptions [(i)-(vi) above] regarding the scattering properties of the medium (and its boundaries) and spatial processing (i.e., beam pattern structure) are invoked. Again, the pertinent second-moment theory is constructed for the set of waveforms emitted from the scatterers and picked up by the receiver. This includes an evaluation of the frequency-time cross correlation between the complex envelope of the original signal and the reverberation. For the most part, narrow-band signals are again assumed, although results for a special class of wideband signals are developed. Various envelope statistics are also determined, with correction terms, from which conditions on the limiting Gaussian (or Rayleigh) statistics can be constructed (cf. Contents).

Chapter VII: "Problems of Detecting Echo Signals in the Presence of Interference": This chapter is largely devoted to the calculation of error probabilities of detection (and probabilities of correct detection decisions) for both coherent (i.e., matched-filter, or correlation detection) and incoherent (or energy detection) reception. It is here, of course, that the various first- and second-moment calculations of Chapters V and VI are used as appropriate parameters in the familiar Gauss, Rayleigh, and Rice distributions, which represent the underlying interference models for the different specific conditions of observation here. The chapter concludes with a short extension of these results to multichannel systems.

Chapter VIII: "Elements of the Theory of Statistical Measurement": The subject of this chapter is the development of statistical estimators of (any of) the random quantities of interest in the sonar model. Specifically, only quadratic estimators are considered (i.e., those involving a mean-square error calculation, or quadratic cost-function, in the language of statistical decision theory [12]). The mean and variance of this class of estimator are examined and optimum forms are determined, based on smoothing time and a "resolution" or sampling interval for the random magnitude under study. An adaptive optimization procedure applied to various parameters of the measuring system is also discussed here.

<u>Chapter IX: "Inverse Probability Problems in Sonar"</u>: This chapter is devoted to various "inverse" problems in sonar: given the injected and received signals, geometries, etc., what can be deduced about the detailed effects of the inhomogeneous medium? To pursue this it is necessary to determine whether such questions represent "well-posed" or "ill-posed" problems, by establishing whether the "correctness conditions" (existence, uniqueness, and stability of solutions) are obeyed. This the author does by applying these concepts to a series of specific "inverse" problems, where it is desired to determine:

(i) the pdf of scatterer Dopplers

(ii) the spectrum of the Doppler fluctuations (which involves an interesting use of Mellin transforms)

(iii) the (relative) level of the coherent component

(iv) the average number of scatterers

(v) the frequency behavior of the scatterers

(vi) the spatial density of the scatterers

[The key conditions underlying one's ability to obtain the specific results here are: (i), the omnidirectionality (and hence frequency insensitivity) of the acoustic arrays coupling to the medium, and (ii), the short duration of the injected signals, to insure ignorable variations with time (i.e., range) over a typical signal duration in the medium; cf. comments below.]

The book concludes with a rather extensive list of unsolved problems (most of which still remain candidates for further investigation at this date) and a variety of recommended steps toward developing a program for the systems engineering of experimental research in the (active) sonar area, where systems of greater complexity can be studied. A selected bibliography and set of references (selectively expanded by us for this edition) completes the book.

The statistical model of reverberation used in this work is that developed earlier in the so-called FOM (Faure [33], Ol'shevskii [14], Middleton [35, 39]) theory, and is thus subject to the limitations of that theory. These are principally: (1), the assumption of independent scattering, with a consequent neglect of multiple scatter effects; and [also stemming from (1)], (2), omission of spatially distributed statistical effects, particularly at boundaries, such as the directional, as well as temporal, correlations of the ocean wave surface. The latter are the main components of the classical, or continuum, "wave models" (for example, [31, 42-44, 64]), which, however, do not yield the independent specular-point or Poissonian component ("speckle" in optical scatter pheonomena) which is the basic scattering mechanism of the FOM models employed here.

Thus, although FOM models appear satisfactory for (weak) volume inhomogeneities, such models are incomplete when interface (wave surface and/or bottom) scattering is involved, since they neglect the random scattering component associated with the continuous portion of the random interfaces, which even at high frequencies can make a noticeable contribution to the total reverberation. On the other hand, the classical continuum and FOM theories can each account (albeit differently in detail) for possible specular (i.e., coherent) components reflected from the interfaces. Moreover, both the FOM and classical "wave models" can be shown to be coexisting subsets of a more general and profound scattering theory, which includes multiple-scattering, intermodal interactions (e.g., surface $\otimes$ volume scatter), and the global spatio-temporal history of the pertinent scattering operators in the associated Langevin equations of the propagating fields, all from a viewpoint which combines the appropriate physical structure (wave and boundary equations, and initial conditions, geometry, etc.) with communication-theoretic concepts and statistical methods; cf. Middleton [63].

Other acknowledged limitations on the scope of the treatment here are the assumption of Gaussian statistics (as the limit of a large number of independent scattering events perceived at the receiver) and the use of omnidirectional acoustic arrays, which are also required to be frequency insensitive. The former is reasonable as long as the number of effective scatterers affecting the receiver is large, and no few, particular scattering elements respond more strongly than the rest to the incident field. When this is not the case, the applicability of a second-moment theory, characteristic of Gaussian processes, breaks down, and (as the author also notes) higher moments of even very high order are required to help specify receiver performance. The latter assumption (of omnidirectional arrays) makes the explicit treatment of inverse problems (given in Chapter IX) analytically practicable.

The usual apertures employed by most sonar systems, however, are directional (and frequency sensitive [69]), and thus, with them, the extraction of scatterer structure and properties involves a spatial deconvolution of the scattered field and these coupling apertures, an analytically nontrivial task. Similar considerations affect the analytic forms of the various injected and received signals [13, 35, 39] generally.

Another limitation is the assumption of zero-gradient media ($\nabla c = 0$), which, while pedagogically no impediment to the basic approach illustrated here, does confine the applicability of the specific results to a rather restricted class of physical situations: high frequency and short ranges, or special ocean situations where $\nabla c = 0$ for whatever frequencies and ranges may be selected [42].

Finally, other important associated topics, such as array processing [13, 40, 41, 48, 60, 69], the specific role of the ocean surface (and bottom) in reverberation [41, 43, 44, 63, 65], and the effects of ambient noise fields, are not considered, in keeping, however, with the author's specified intent to illustrate the use of statistical methods in active sonar in a work of convenient and moderate size.

Apart from the clear exposition and organization, and in addition to the specific results of Chapters VI and VII (particularly those involving "noise" signals), there are a number of highlights which should be of particular interest to the reader. These appear for the most part as the topics of Chapters VIII and IX. Also, the concluding section offers an interesting (and still largely unachieved) program of pertinent future studies. As a useful exposition of the statistical approach in this field, the book fulfills its main purpose, along with a variety of specific results not readily available in book form elsewhere.

The editor (D.M.) has modified the translation with only minor changes, principally to adjust terminology here and there to Western usage (e.g., "ambiguity functions" for "indeterminacy functions," "quantity" or "estimate" for "valuation," etc.); he has corrected a number of obvious misprints, and has incorporated the author's recent (1976) corrections to the original Russian edition. Besides this Foreword, and a number of explanatory footnotes throughout the text, the editor has added a selected set of pertinent references, which were not available to the author at the time, and which it is hoped will further enhance the book's interest and utility.

David Middleton

New York, October 1977

PREFACE TO THE AMERICAN EDITION

This book was conceived as an illustration of the use of statistical methods in active sonar, which is one of the effective ways of exploring the ocean using statistical methods. In keeping with the restricted scope of the book, only isolated questions in the statistical theory of sonar were considered. Undoubtedly the body of problems which might be successfully resolved by means of statistical methods is significantly broader, and a great deal remains to be done.

The author is indebted to Plenum Publishing Corporation for the interest shown in this book, the more so because it is the second book by the author to be issued in the USA by this publisher (the monograph "Characteristics of Sea Reverberation" was published in 1967).

I consider it a pleasant duty to express my gratitude to the initiator of the present publication -- David Middleton. Personal contact with Professor Middleton has not only been useful in a scientific sense, but also has given me great pleasure.

V. V. Ol'shevskii

Moscow, March 1976

PREFACE TO THE RUSSIAN EDITION

Sonar methods are presently enjoying increasingly greater application. For example, active sonar systems are being used to study various physical characteristics of the oceans, seas, and river basins, to search for minerals located under water, to search for fish and other marine animals, and so on. Such diverse, effective uses of sonar systems demand considerable development of the theory and the engineering behind these systems, chiefly the development of probability and statistical methods in sonar.

This book is devoted to a presentation of selected problems in the use of probability and statistical methods in sonar. These methods play a decisive role in the formulation and conduct of quantitative experimental investigations for the purpose of comparing statistical measurements with probability models of the objects being studied.

The contents of the book is presented in the following order.

The introductory chapter, Chapter I, discusses the sonar model and its principal statistical aspects.

Chapter II is devoted to the development of quantitative methods for assessing the quality of experimental sonar research. Attention is turned to the need for developing probability models of the subjects of research and for making statistical measurements in experiments.

Various methods for constructing probability models of sonar signals, by which the subjects of research can be described, are examined in Chapter III. Most of the attention is devoted to the development of phenomenological probability models -- canonical, parametric, and constructive.

Chapter IV contains a classification and review of most of the emitted signals encountered in practice.

The characteristics of echo signals satisfying various probability models of distortion stemming both from reflection of sound waves by objects being detected and from the effects of the conditions imposed by underwater observation are examined in Chapter V. The characteristics of echo signals are compared through spectrum correlation.

Chapter VI presents a probability analysis of reverberatory signals for the case in which acoustic waves are scattered by discrete irregularities in the water medium.

Chapter VII is devoted to a study of the detection of echo signals in the presence of interference. Information processing methods typical of sonar are analyzed.

Chapter VIII presents some concepts of modern statistical measurement theory as a means for obtaining quantitative data on the probability characteristics of random processes. Methods for optimizing statistical measurements are also discussed here.

Various inverse probability problems encountered in sonar, which involve determining the properties of scatterers on the basis of the structure of received signals, are examined in Chapter IX.

The author confesses that the book offered to the reader does not include some interesting, pressing problems of statistical theory and sonar. Its chief purpose is to illustrate the effectiveness of a statistical approach for solving some important problems of both theoretical and practical importance.

The author is grateful to the reviewers of the book, V. I. Klyachkin and A. K. Novikov, for their useful remarks, and to T. P. Zlobina and V. A. Pulle for their assistance in preparing the manuscript.

CONTENTS

CHAPTER I

INTRODUCTORY REMARKS

§1. Some Definitions

Two basic hydroacoustic methods for underwater observation are presently being developed -- passive and active sonar.

Passive sonar* is a method for establishing the presence and determining the properties of underwater objects through the reception and processing of acoustic signals emitted by these objects.

Active sonar is a method for establishing the presence and determining the properties of underwater objects based on injecting hydroacoustic signals into the water medium and then receiving and processing the received echo signals arising when these sound waves are reflected (or scattered) by underwater objects.

It is possible to use passive sonar as a method of underwater observation only in the event that the underwater object of interest emits acoustic signals -- that is, when it is a source of a primary hydroacoustic field, while active sonar can be used all the time in principle, in view of the fact that an echo signal -- that is, a secondary hydroacoustic field -- carries the information about the underwater object in this case.

We will subsequently employ the term "sonar" to mean "active sonar" whenever the implication of the term remains obvious.

---

*Sometimes the term "hydrophonic detection" [shumopelengovanie] is used in place of this term. However, it does not fully reflect the functions of passive sonar as formulated here.

We define subjects of detection, or subjects of research, as underwater objects of artificial or natural origin from which acoustic waves are reflected (or scattered) -- that is, those which form echo signals. Studying the properties of the latter is the task of sonar.

The basic tasks of underwater observation that can be fulfilled by sonar systems include detection of objects, measurement of their current location, and their classification.

Detection is defined as establishing the presence (or absence) of the desired object of detection within a certain volume of the water medium, which can be rather large. Thus, the answer obtained in response to execution of detection tasks has a binary nature: "The object is present" or "the object is absent."

Measurement of current coordinates (location and motion) is defined as obtaining quantitative data on the coordinates of the object of detection as a function of time.

Classification is defined as identifying the type of object of detection -- that is, its identification with one of several classes of interest to us.

A sonar system (station) is a complex of devices and mechanisms used in making underwater observations by sonar methods. A sonar system normally includes:

an input generator, which produces the emitted signals;

a sound-radiating antenna, for the purpose of forming the acoustic field of the emitted signal;

a sound-receiving antenna, for the purpose of receiving and processing the acoustic field of echo signals;

rotating or switching devices operating on the radiating and receiving antennas such that a water medium sector is formed and scanned;

received signal amplifiers;

a device to process inputs for detection tasks, measuring current coordinates, and classifing objects of detection;

displays by which the results of echo signal processing are presented to the sonar operator;

a sonar system control device;

streamlined acoustic antenna housings.

§2. The Sonar Model

When combined with the water medium and the objects of detection -- that is, the objects of research, the sonar system, which has the purpose of obtaining various types of information on the properties of these objects, can be treated as an information system. Examination of the characteristics of sonar systems from the information viewpoint is useful to gaining an understanding of the role of different factors affecting the characteristics of these systems. Our attention is essentially focused on the statistical aspects of sonar in such an examination. The sonar information system has been discussed in [13,33,35] and is similar to a considerable degree to the information system applicable to radar systems [6,7,26,32]. There is also a certain analogy between this system and the information system of communications [1,12,45,46,47,51].

Let us examine the information system of sonar (Figure 1).

The source of emitted signals is defined as the combination of resources (input generator, power amplifiers, rotating and switching devices, and sound-radiating antenna) with which we generate the acoustic field of the emitted signal $C(t,\varrho)$ in the water medium, where t is time and $\varrho = (x,y,z)$ are spatial coordinates. In this case if we consider $C(t)$ to be an electric signal that excites the radiating antenna, then

$$C(t,\ \varrho) = AC(t), \tag{2.1}$$

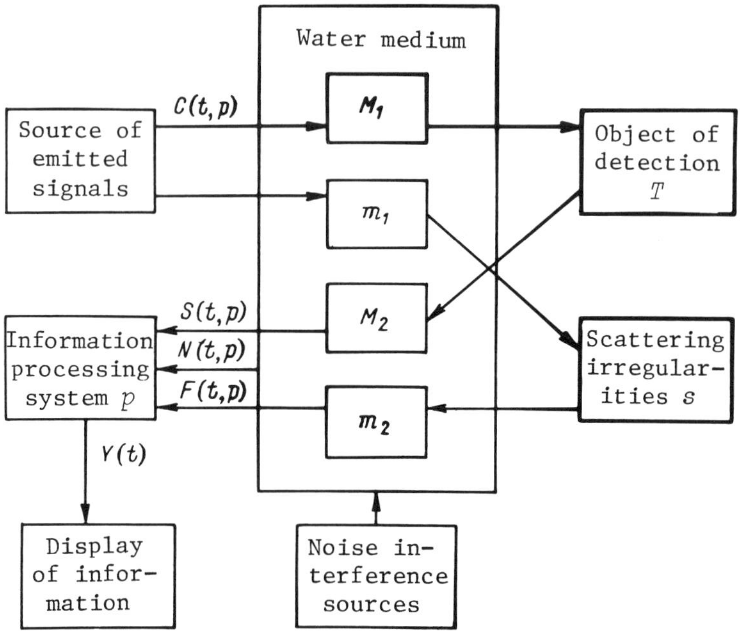

Figure 1. The sonar information system.

where A is the operator* forming the emitted signal's acoustic field. Signal $C(t,\varrho)$ propagates through the water medium -- that is, it passes through a communication channel, and is subjected to distortion within it. The causes of the distortion of the signal $C(t,\varrho)$ as it propagates are:

(i) refraction phenomena stemming from the relation between the mean rate of propagation of acoustic waves and the spatial coordinates;

(ii) reflection and scattering of acoustic waves from water-surface boundaries and from the bottom;

(iii) temporal variability of the water medium's acoustic characteristics and of the waviness of the surface.

*Equivalently, the transmitting aperture's weighting function. [D.M.]

As a result we have an acoustic field $C_{M_1}(t, \varrho)$ at a certain point in the water medium, related to $C(t, \varrho)$ by the equation

$$C_{M_1}(t, \varrho) = M_1 C(t, \varrho) = M_1 AC(t), \tag{2.2}$$

where M_1 is an operator taking account of the effects of factors (i) - (iii) enumerated above.

The object of detection reflects and scatters signal $C_{M_1}(t, \varrho)$ such that the echo signal $S_{M_1}(t, \varrho)$ observed near the object of detection could be given in the form

$$S_{M_1 T}(t, \varrho) = TC_{M_1}(t, \varrho) = TM_1 AC(t), \tag{2.3}$$

where T is an operator taking account of the object's properties (its shape, length in space, movement characteristics). At the point of reception the echo signal is described by the function $S(t, \varrho)$, which is related to $S_{M_1 T}(t, \varrho)$ by operator M_2, which takes account of distortions stemming from propagation of the echo signal from the object of detection to the point of reception, such that

$$S(t, \varrho) = M_2 S_{M_1 T}(t, \varrho) = M_2 TM_1 AC(t). \tag{2.4}$$

We note that operators M_1 and M_2 may be identical when the sonar model is examined simplistically. However, in the typical cases of long-range sonar, in which the spatial and temporal properties of the water medium and its boundaries must be taken into account, operators M_1 and M_2 are different.

In addition to echo signals, various types of interference pass into the reception area.

Reverberatory interference arises when the emitted signal is scattered by irregularities in the water medium and unevennesses at its boundaries. Within the reception area the reverberation acoustic field $F(t, \varrho)$ is defined as

$$F(t, \varrho) = m_2 s m_1 AC(t), \tag{2.5}$$

where m_1 and m_2 are operators taking account of the effects of the propagation conditions (operator m_1 takes account of distortions of primary signals as they propagate from the radiating antenna to scattering irregularities, while operator m_2 takes account of distortions of scattered signals as they propagate from the irregularities to the reception area), and s is the operator describing the spatial and temporal properties of the scattering irregularities.

Interference of natural origin appears as a consequence of waves on the water surface and the vital activities of marine animals which emit acoustic signals. We designate the field of this interference in the reception area by $N_1(t, \varrho)$.

Sonar system platforms generate noise interference as a result of:

(a) the working of various devices and mechanisms installed on the transporters -- vibration interference* $N_v(t, \varrho)$;

(b) the effects of pulsations in hydrodynamic pressure on the hull -- hydrodynamic interference $N_h(t, \varrho)$;

(c) development of noise during cavitation -- cavitation interference $N_c(t, \varrho)$.

To sum up, the acoustic field of noise interference $N_2(t, \varrho)$ generated by the sonar system platforms can be presented in the form

$$N_2(t, \varrho) = N_v(t, \varrho) + N_h(t, \varrho) + N_c(t, \varrho). \qquad (2.6)$$

Another type of noise interference, $N_3(t, \varrho)$, stems from acoustic emissions produced by objects of detection as a result of, for example, their movement.

---

*In this case "vibration interference" is an abbreviated rather than a precise term. It would be more accurate to refer to acoustic interference having origins in vibration, one component of which spreads through the hull of the sonar system's platform while the other spreads directly through the water medium.

To sum up, we observe an acoustic field $X(t, \varrho)$ in the reception area, representing a mixture of the echo signal and different types of interference:

$$X(t, \varrho) = S(t, \varrho) + F(t, \varrho) + N_1(t, \varrho) + N_2(t, \varrho) + N_3(t, \varrho), \quad (2.7)$$

or, in expanded form, taking account of expressions (2.4)-(2.6),

$$X(t, \varrho) = M_2TM_1AC(t) + m_2sm_1AC(t) + \\ + N_1(t, \varrho) + N_v(t, \varrho) + N_h(t, \varrho) + \\ + N_c(t, \varrho) + N_3(t, \varrho). \quad (2.8)$$

The acoustic field $X(t, \varrho)$ passes into the information processing system, which contains the sound-receiving antenna, the collection of linear and nonlinear processing devices, and the displays. As a result, we get the following process at the processing system's output:

$$Y(t) = PX(t, \varrho), \quad (2.9)$$

where P is the processing operator. Its form depends on the specific nature of the task being executed (detection, measurement of current coordinates, classification).

Next, data on $Y(t)$ are used by the receiver of the information, which could be either the sonar operator or a certain technical device. The functions of the latter include developing and making particular decisions on the basis of $Y(t)$.

Using information available on the conditions of underwater observation, the sonar operator controls the operating conditions of the sonar system (independently or in association with some sort of automatic machine).

Thus, the sonar system should be viewed as a complex information system.* One group of this system's components is defined by the conditions of underwater observations:

---

*See [13], for example, Figure 1 and Sections I, IIA, III for details of coupling to the medium; also [60]. [D.M.]

The characteristics of components in this group cannot be changed willfully by selecting particular sonar system parameters. The second group of components possesses characteristics dependent on the sonar system's parameters. These characteristics can, in principle, be changed within rather wide limits. The first group of components includes the operators M_1, m_1, M_2, m_2, T, and S, which depend correspondingly on properties of the water medium, the object of detection, and the scattering irregularities. This group also includes noise interference $N_1(t,\varrho)$, $N_2(t,\varrho)$, and $N_3(t,\varrho)$. The second group of components includes signal $C(t)$ and operators A and P, which are defined by characteristics of the acoustic antennas and the information processing algorithms. Reverberatory interference $F(t)$ is also related to this group to some degree.

A physical interpretation can easily be given to the operator form for representing sonar information and the conditions for using the corresponding systems. We will dwell on this problem in Chapters III, V, VI, and IX.

§3. Sonar and Radar

Sonar has much in common with radar from the viewpoint of the methods for executing tasks connected with determining particular properties of the objects of detection [6, 22,26]. In fact, the means for obtaining information on the properties of the objects of detection are the same in both cases, and it is the echo signal, the various characteristics of which bear information on the characteristics of the objects of detection, which is subjected to processing. Thus, the tasks of both sonar and radar can be summarized in a general way as follows: Characteristics of the object of detection (research) of interest to us must be determined on the basis of measured (that is, given) characteristics of the echo signal.

In addition, a statistical theory of radar, which has been developed intensively in the last two decades [6,7,26, 32,38,46], has produced a number of general results that can be applied far beyond the bounds of this field. In principle, many of these results can be used directly in a statistical theory of sonar.

The following question arises: Is there anything in

a statistical theory of sonar that is unique when compared to the statistical theory of radar? Were the answer negative, we would naturally assume that a statistical theory of sonar is not independent and that its development would simply involve application of a number of results obtained from the statistical theory of radar. But if the answer is positive, we must view a statistical theory of sonar as an independent scientific entity applicable to at least some problems that have unique traits.

We hold to the second point of view. Differences between statistical theories of sonar and radar appear in several ways, connected primarily with differences in the observation conditions of the two techniques. We will examine some features of sonar below in the light of the information system behind the sonar system, as discussed in §2.

In the first place, the great variability of conditions behind underwater observation is typical of sonar. The reason for this is the spatial and temporal irregularity of the water medium's acoustic characteristics, connected with variability of the conditions under which acoustic waves propagate. As a consequence, operators M_1 and M_2, which in accordance with formulas (2.2)-(2.4) define the distortion experienced by the echo signal (stemming from the effects of propagation conditions), can have an extremely complex form and yield to deterministic prediction poorly. In all cases of practical interest these operators are random, as a consequence of which they can be described and predicted only by statistical methods. This situation is not typical of radar, as is noted in [13,22,26,50,51,60].

In the second place, the objects of research in sonar as a rule have a complex shape, they differ in their length in space, and the movement of these objects (or of their individual parts) follows a complex law. Hence it follows that operator T, which in accordance with (2.3) defines the acoustical (spatial-temporal reflecting and scattering) properties of the object of detection, can transform the emitted signal considerably, subjecting it to both linear and nonlinear transformations. It should be noted that in contrast to the situation in radar, where the effects of operator T are largely limited to changing the intensity, the moment of arrival, and the Doppler shift of the useful signal [6,26,32,65], in sonar the effects of operator T create a situation in which the echo signal assumes a random form (partially or completely). Echo signals can consist of several elementary echo signals, and they may possess additive or multiplicative components, which are expressed to varying degrees due to the way elongated objects of detection move in many cases (for example, when sonar is applied to schools of fish, the models and characteristics of echo signals depend on the properties of these objects).

In the third place, scattering by irregularities in the water medium and by unevennesses at its boundaries causes a unique type of interference in sonar -- marine reverberation. It follows from (2.5) that in this case the characteristics of the interference depend on both the type of signal emitted and the properties of operators m_1, m_2, and s, which are governed correspondingly by the acoustic wave propagation conditions and the properties of scattering irregularities. It is true that interference similar to reverberation, in a particular sense, is present in radar. It is called the local object background* [13,26]. However, the properties of such interference depend to a considerable degree only on the type of signal emitted and the characteristics of the scattering objects, while according to (2.5) reverberatory interference also depends on properties of the water medium.

In the fourth place, the low velocity of propagation of acoustic waves, as well as restriction to low frequencies lying primarily within the acoustic band, prevents us from gaining any significant quantity of information on the objects of detection with sonar systems (as compared, of course, with the use of radar systems in air under equivalent conditions).

In the fifth place, in sonar the strongly pronounced Doppler effect experienced by echo signals connected with movement of the objects of detection leads to the necessity of developing echo signal models in which the Doppler effect influences the shape of the echo signal and does not simply shift its central frequency [13,22,65].

---

*For example, "clutter" or "ground clutter." [D.M.]

The features of sonar system use enumerated above indicate that a statistical theory of sonar contains a whole series of problems of a probabilistic and informational nature which are unique and which hold independent interest.

§4. Statistical Problems of Sonar

Intensive development of sonar equipment, expansion of its area of application, and, finally, the increasing complexity of the problems that must be solved with sonar systems have all led to the need for developing statistical methods for describing the systems and the conditions in which they are used. Accordingly, the problem has arisen of developing a statistical sonar theory [13,14,17,19,22, 33-38,60] in the framework of which we can establish general laws applicable to the use of sonar systems of different designs under different conditions, define quality criteria for the systems, and develop methods for optimizing their parameters and characteristics.

Such a theory must be statistical for the following reasons. In the first place, by their nature sonar problems are probabilistic, since the volume of *a priori* information on the objects of research and the conditions of underwater observation required for their solution is always limited -- that is, some of the properties of these objects can be described only by statistical methods. In the second place, signals and interference applicable to sonar are described by random processes or random fields. In view of these reasons the characteristics of sonar systems can be analyzed by statistical methods, and quality criteria for sonar systems can be developed on the basis of data relevant to the applications of these systems, the statistical properties of signals and interference, and the conditions of underwater observation. This is why probability methods are used to solve the problems of optimizing parameters and characteristics of sonar systems.

Statistical theory of sonar is that division of hydroacoustic engineering concerned with the development of probability models for signals, interference, and underwater observation conditions and, on the basis of these models, development of methods for analyzing and synthesizing sonar systems. The development of a statistical theory of sonar is intimately associated with the development of other informational and probabilistic directions of science and engineering. Here we should first note a statistical theory of communication (or general theory of communication, as it is sometimes called) and a statistical theory of radar [6,12,26,32], which has been concerned with developing methods for analyzing and synthesizing engineering systems (radio systems primarily) designed to transmit, receive, and process useful signals on the basis of probability theory, mathematical statistics, and random functions theory.

Let us examine the content of a statistical theory of sonar.

The problem -- that of obtaining *a priori* information -- includes studying the conditions under which acoustic signals propagate through a water medium, the characteristics of signal reflection from objects of detection, the characteristics of signal scattering from irregularities in the water medium and at its boundaries, and the mechanisms by which various types of noise interference arise. We obtain this information by conducting research on these characteristics. In this case the volume of research and the combination of conditions studied must be such that the data obtained are adequately representative and reliable. This requires that the research be carried out in two stages (assuming we desire to approach the search for and use of *a priori* information correctly with respect to methodology). At the first stage we must determine some of the random statistical characteristics of the functions under investigation. For this purpose it would be sensible to make use of theoretical research methods primarily (in some cases when a theoretical solution to a problem is excessively complex we can also employ modeling methods). The correctness of the obtained results could be tested by performing suitably designed experiments. It also becomes necessary to perform experiments at this stage when it appears impossible to formulate the task of theoretical research precisely or to define the initial data for modeling purposes.

Experiments must be based on statistical measurements -- that is, on determining the values of statistical characteristics of interest to us with minimum and known (controllable) error [53, 58,62]. Otherwise, when experiments are reduced to obtaining qualitative results, it is normally impossible to draw conclusions concerning the adequacy of models and the object of research, even upon careful examination.

In the second stage of obtaining *a priori* information we study the effects of conditions (initial data) on the statistical characteristics of the functions under investigation. At this time we determine the unconditional statistical characteristics of primary, reflected, scattered, and interference signals which, when combined with the random characteristics, make up our *a priori* information.

In the last decade we have accumulated an extremely large amount of information from research on properties of the water medium, objects of detection, irregularities, and sources of underwater noise [2,4,13-17,20-22,23,28,30,31, 33-36,41,42,48,53,56-59,62].

Probability models of signals, interference, and the conditions of underwater observation are being developed on the basis of the *a priori* information obtained. In this case we encounter two manifestations of randomness: Functions describing models of echo signals and interference and functions describing the underwater observation conditions are random. This forces us to develop probability models of signals and interference in two stages. In the first stage we develop conditional models of signals and interference -- that is, models related to certain set conditions of underwater observation. At the same time models of underwater observation conditions are developed. In the second stage we study the effect of the underwater observation conditions on the models of signals and interference, and consequently we develop their unconditional probability models.

It stands to reason that models describing signals, interference, and the conditions of underwater observation cannot be developed with adequate fullness just on the basis of available *a priori* information alone. As we develop such models we must obtain additional information with which we can correct and improve the models such that they begin to approximate the objects of research. There is presently a large number of probability models describing signals, interference, and underwater observation conditions [2,13,14,19-22,23,31,33-36,39,41,42,48,52,59,62,63,65]. However, this problem is still in its initial stage of development, since in many cases we have not yet been able to establish the adequacy with which the models describe real phenomena.

The next problem is that of obtaining current information. We can view the sonar system as a statistical measuring system. We can use this system to measure some probability characteristics of signals and interference observed under specific conditions. These characteristics can change with respect to time and spatial coordinates. Consequently we must talk in terms of measuring current characteristics that comprise current information.

Here, as in the case of obtaining *a priori* information, our attention is focused on the measurement aspect of the problem.

Experiments must produce statistical values for probability characteristics with a known degree of precision. Obviously when current information is used, models of signals, interference, and underwater observation conditions would be more precise than when only *a priori* information is employed. We should note the considerable amount of progress experienced in the theory and methods of statistical measurement in the last few years [3,8,9,11,18,20,21,23,24,53,56-58,60-62]. In particular, questions have been raised involving the proper way to measure the characteristics of random processes and fields, ways to optimize the measurements, and the measuring tools necessary to support such measurement.

Let us now turn to a brief review of the tasks involved in analyzing and synthesizing sonar systems.

The central task is that of developing indices of effectiveness for sonar systems. At least the following three factors must be considered in this regard:

1. the goals to be attained through the use of sonar systems;

2. the essential parameters or characteristics of the sonar systems;

3. the characteristics of signals, interference, and of the underwater observation conditions.

Indices of effectiveness are normally some sort of functionals obtained by statistically averaging conditional functionals with respect to those parameters or characteristics that are random. The following characteristics of sonar systems must be taken into account in defining these indices [5,7]: the probability that the tasks would be performed successfully, the time required to make a decision the water medium volume subjected to observation (the scanning sectors, the ranges), economic expenditures, energy expenditures, and the weight and size parameters of the system.

Indices of effectiveness can be particular or general. When developing particular indices we consider just the individual (particular) tasks performed by sonar systems and the expenditures necessary for such performance. The general index of effectiveness is composed of particular indices, and the algorithm for its composition must be adequately justified.

The problem of analyzing sonar systems of a certain design is reduced to examining their operation under different conditions of underwater observation. In this case the probability distributions for variations in these conditions are determined on the basis of *a priori* information and the models developed. The analysis should produce indices of effectiveness for the sonar systems in quantitative form so that we could use them to compare different systems, reveal the causes of unsatisfactory operation, define the conditions under which the system operates with a particular degree of effectiveness, and so on.

The problem of synthesizing sonar systems, as one part of a statistical theory of sonar, involves selecting a version of a system that is best in a certain sense. In other words synthesis of sonar systems can be reduced to a system optimization problem.

In concluding this examination of the statistical problems of sonar we emphasize once again that this examination is based on probability models describing signals, interference, and underwater observation conditions specific to sonar. Improving such models, testing the adequacy with which they reflect the real objects of research using statistical measurements, and solving inverse problems of sonar are the key problems examined in this book.

CHAPTER II

PROCEDURAL FUNDAMENTALS OF EXPERIMENTAL RESEARCH IN SONAR

§5. Principal Tasks of Experimental Research

Sonar experiments can be conducted at qualitative and quantitative levels. In qualitative experiments the researcher obtains a certain characteristic of the hydroacoustic signal for which the measurement accuracy is unknown. Nor, consequently, does he know its relationship to the true characteristic. In this case the main goal of the experiment is to establish a particular property for the object of research (for example, to establish that the object of detection moves, that scattering off of irregularities is occurring and that its intensity exceeds that of noise occurring naturally, that there is an echo signal in the input process, etc.).

Circumstances are different in quantitative experiments. The principal feature of such experiments is that of obtaining the hydroacoustic signal characteristic under investigation with a known degree of accuracy -- that is, with controllable measuring error. In this case the purpose of the experiment is to measure precisely the characteristics of interest -- that is, to obtain values for them (for example, to measure the probability distribution of the hydroacoustic signal, its moment functions, its spectra, etc.).

The remainder of this chapter is based on information in [8, 9, 18, 20, 21, 23, 24].

We can define the following four principal tasks performed in experimental research in sonar:

1. accumulation of information regarding known properties of the object of research;

2. exploration for new properties of the objects of research;

3. testing the validity of a theory (model) developed;

4. solution of inverse problems.

The first two tasks are usually performed at the qualitative level while the third and fourth are performed at the quantitative.

The volume of formalized *a priori* information available to the researcher for planning and conducting experiments is an important factor, as it affects their nature. For example, in qualitative experiments the volume of *a priori* information is ordinarily so small that it is impossible to achieve a satisfactory mathematical description of the object of research and the hydroacoustic signals corresponding to it. In quantitative experiments the volume of *a priori* information must be adequate for mathematical description of the object of research. When this condition is observed it becomes possible to shift from qualitative to quantitative experiments.

We cannot, however, ignore typical situations in which the *a priori* information available is used incorrectly by investigators or not used at all. In such cases experiments that could have produced quantitative (that is, more valuable) results in principle do in fact produce only qualitative (that is, less valuable) results. In addition to improper general procedures used in setting up and conducting such experiments, we also encounter inefficient expenditure of material resources and high research cost.

We feel that almost all types of experimental research in sonar can be conducted on a quantitative level today because a rather large *a priori* base applicable to the development of the most diverse types of objects of research does exist. Thus, the success of experimental research depends on selection of procedures for setting up and conducting experiments that take account of all *a priori* information.

§6. The Structural Diagram of Quantitative Experimental Research

Figure 2 shows a structural diagram of quantitative experimental research in sonar reflecting the general procedures for setting up experiments. We will begin examining this diagram using hypothetical objects of research, or models of real objects. The hypothetical objects are developed on the basis of formalized *a priori* information a on the properties of the real objects available to the investigator, and the hypotheses h which the investigator feels he must assume. Both taken together make up the initial data. There is a fundamental importance in distinguishing between the formalized *a priori* information and the hypotheses composing the initial data used to develop

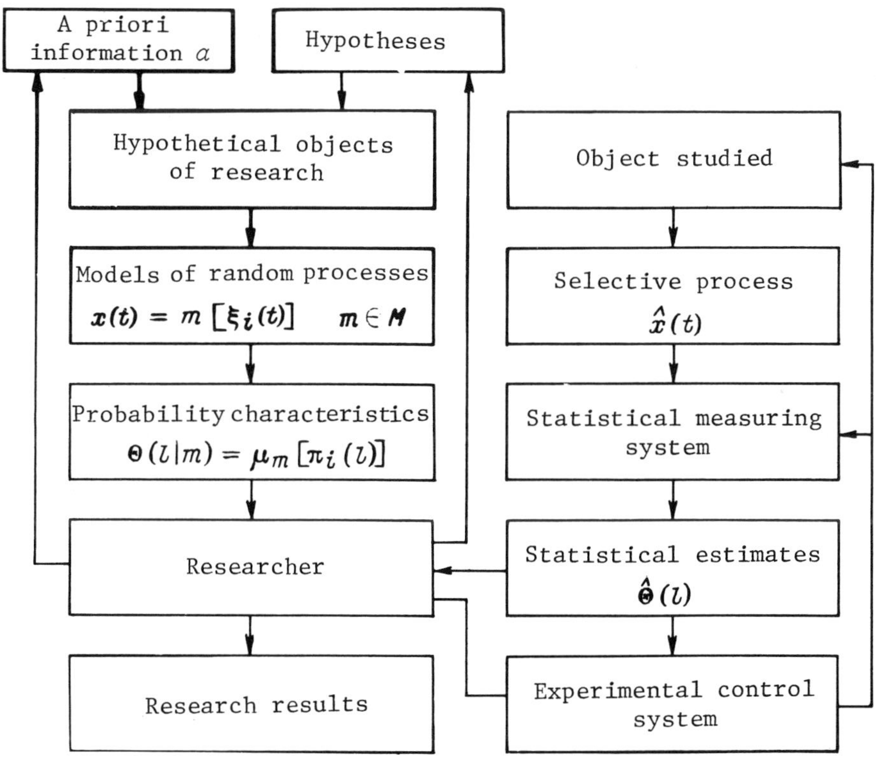

Figure 2. Structural diagram of experimental research.

models of the objects. Formalized *a priori* information has the experience of previous research at its basis. These data are reliable or almost reliable. Meanwhile, hypotheses are based on the guesses of the investigator, analogies with other branches of science, and intuitive deliberations. These data are not reliable and must be tested experimentally.

The list of hypothetical objects must be complete enough so that it embraces all possible modifications of the real object and the conditions under which this object may exist.

When using the methods of sonar research, we study not the objects themselves but rather some functions -- acoustic fields or temporal processes -- that describe their properties. Random fields $x(t,\mathbf{Q})$ or random processes $x(t)$, are the idealized models of such functions. We will examine random processes below on the assumption that applying the results of this discussion to the case of fields does not present difficulties.

As we know, a random process $x(t)$ is defined by an infinite set of time functions $x_k(t)$, $k = \pm 1, \pm 2, \ldots, \pm \infty$, each of which is associated with one of the hypothetical observations and is termed representation of a random process. The properties themselves of a random process are described by probability distributions or by their parameters (moment functions, semi-invariates, etc.). Thus, we assume that the characteristics of the hypothetical objects are reflected by random processes corresponding to them, and the properties of these processes are defined by formalized *a priori* information and the hypotheses adopted.

The probability model, which is defined as any sort of representation of a random process enabling us to compute or postulate its probability characteristics* which are significant to the problem being solved, is a convenient (and perhaps the only) form for taking account of *a priori* information a and hypotheses h. In this case the random

---

*Equivalently, *statistics*. (D.M.)

process of interest $x(t)$ is represented in the following form:

$$x(t) = m\{\xi_i(t)\}, \tag{6.1}$$

where $m = (a,h)$ is an operator characterizing the type of model selected on the basis of *a priori* information a and hypotheses h; $\xi_i(t)$, $i = 1,2,\ldots,p$ are random processes for which the probability properties are assumed to be known. Models $m \in M$ must form set M that describes the properties of the real objects of research with adequate fullness (the symbol $\in$ designates that model m belongs to set M).

Let $\Theta(l|m)$ be the probability characteristic* of the random process $x(t)$ corresponding to its model m, where $l = (l_1,l_2,\ldots,l_n)$ is the set of independent variables appearing as arguments of the characteristic under examination (n defines its dimensionality). Obviously, the characteristic $\Theta(l|m)$ is conditional from the point of view of its dependence on model m. Then let $\pi_i(\lambda)$, $i = 1,2,\ldots,p$ be the probability characteristics of random processes $\xi_i(t)$ and $\lambda = (\lambda_1,\lambda_2,\ldots,\lambda_r)$ be the set of corresponding independent variables.

By definition, the probability model m must enable us to establish a connection between the initial (known) probability characteristics $\{\pi_i(\lambda)\}$ and the characteristic $\Theta(l|m)$. In other words, setting up a probability model means defining the equation

$$\Theta(l|m) = \mu_m\{\pi_i(\lambda)\}, \tag{6.2}$$

which follows from expression (6.1). In this case operator μ_m must be defined uniquely by operator m -- that is, by the model of the random process.

It should be kept in mind that for sufficiently complex objects of research and conditions under which they are observed, equation (6.2) may turn out to be a dynamic equation. This means, in particular, that operators μ_m and m depend on time. Such a situation requires that we formulate additional

---

*Equivalently, a *statistic*. (D.M.)

conditions which must be satisfied by the probability model, and refine the definition of the measure of probability on which construction of the model is based. The principle of homogeneity of the objects of research and the conditions under which they are observed must in this case be substantiated properly on the basis of the available *a priori* information.

Thus, the investigator gains possession of a collection of characteristics $\Theta(l|m)$, each of which is defined by one of the models m. The investigator need not be defined necessarily as one particular individual (for example, the so-called experimenter). We can refer to a group of individuals -- a scientific collective -- who, in combination with the relevant technical resources, plan the experimental research, conduct it, and interpret the results.

The object of study (or the real object) provides us the possibility for extracting experimental results. In conducting sonar experiments we obtain a certain set of time functions $\{\hat{x}_q(t)\}$, $q = 1,2,\ldots,N$, each of which can be defined as a selective representation and corresponds to the qth experiment on the object of study. Meanwhile the entire set $\{\hat{x}_q(t)\}$ of these representation forms the selective process $\hat{x}(t)$.

We note that there is a difference between the random process $x(t)$ and the selective process $\hat{x}(t)$, and between the representation $x_k(t)$ of process $x(t)$ and the selective representation $\hat{x}_q(t)$ of process $\hat{x}(t)$.

On one hand, inasmuch as $x_k(t)$ and $x(t)$ are associated with a probability model, their properties -- that is, their probability characteristics -- are known, at least in a conditional way (that is, under the condition that a certain model m is adopted). At the same time, the properties of the selective representations $\hat{x}_q(t)$ and of the selective process $\hat{x}(t)$ are not known beforehand (prior to the experiment).

On the other hand, all values of $\hat{x}_q(t)$ and correspondingly $\hat{x}(t)$ become known *a posteriori* (after the experiment) because these functions can be recorded with the needed degree of accuracy, fed into a memory, and so on. In other words, no elements of randomness exist *a posteriori* in the selective representation and selective processes.

These features of random processes $x(t)$ and selective processes $\hat{x}(t)$ stem from the fact that $x(t)$ is always a probability model, the result of introducing a hypothetical object of research, while $\hat{x}(t)$ is the result of a physical experiment on a real object of research.

According to the structural diagram of experimental research we are examining, the selective process $\hat{x}(t)$ is fed into the statistical measuring system, which has as its output a statistical estimate $\hat{\Theta}(l)$ of the probability characteristic $\Theta(l|m)$ of interest to us. We should take special consideration of the meaning which statistical processing of the selective process $\hat{x}(t)$ has if after the experiment its values become known and are not random. The answer to this question involves the nature of the models of the objects of research being developed. As soon as the model of the object of research becomes a probability model, and when $x(t)$ is assumed to be a random process, the selective process $\hat{x}(t)$ should be interpreted as an experimental analog of $x(t)$. All of this means, in turn, that the statistical estimate $\hat{\Theta}(l)$ of the probability characteristic $\Theta(l|m)$ should be the result of the experiment, since it would be natural to consider the methods for describing $x(t)$ and $\hat{x}(t)$ to be the same.

And so, the investigator has available a set of computed (or postulated) probability characteristics $\Theta(l|m)$ and the statistical estimates $\hat{\Theta}(l)$ of these characteristics, obtained by experiment. In the end, the purpose of the experimental research is to compare $\hat{\Theta}(l)$ and $\Theta(l|m)$ and establish a probability model m that is closest to reality.

§7. Discrimination Functional for the Probability Characteristic and Its Statistical Estimate

In quantitative experiments we must determine the numerical correspondence between the probability characteristic $\Theta(l|m)$ and the statistical estimate $\hat{\Theta}(l)$ of this characteristic. In essence we cannot obtain a quantitative experimental result without determining this correspondence.

We introduce discrimination functional $\rho_\Theta(m)$ for the characteristics of interest:

$$\rho_\Theta(m) = \rho\,[\Theta(l \mid m),\ \hat{\Theta}(l)], \tag{7.1}$$

in which case ρ is an operator designating the form in which this difference is defined.

Thus $\rho_\Theta(m)$ is a number defining in a certain sense the distance between functions $\Theta(l \mid m)$ and $\hat{\Theta}(l)$ in the space described by their arguments l.

Here are some general properties of the discrimination function $\rho_\Theta(m)$:

(a) $\rho_\Theta(m) = 0$ only in the event that $\Theta(l \mid m) = \hat{\Theta}(l)$ -- that is, when the characteristics match;

(b) if $\Theta(l \mid m) \neq \hat{\Theta}(l)$, then $\rho_\Theta(m) > 0$;

(c) as the difference between $\Theta(l \mid m)$ and $\hat{\Theta}(l)$ increases, the value of the functional $\rho_\Theta(m)$ should grow.

Introduction of the discrimination functional enables us to determine the error of statistical measurement and to present the experimental results in quantitative form using numerical evaluations.

The mean square functional is a typical discrimination functional. The method of averaging the quadratic function $[\Theta(l \mid m) - \hat{\Theta}_q(l)]^2$, which describes the difference between measurements of $\Theta_q(l)$ and the estimate $\hat{\Theta}_q(l)$, plays an important role in the formulation of such a functional. We will examine two methods for averaging this quadratic function, below.

Let $\hat{\Theta}_q(l)$ be a statistical estimate obtained for the qth selective representation $\hat{x}_q(t)$. The mean square discrimination functional of the first type has the following form:

$$\rho_\Theta(m) = \sum_q A(q) \int_l A(l)\,[\Theta(l \mid m) - \hat{\Theta}_q(l)]^2\,dl, \tag{7.2}$$

in which case summation is carried out over the needed volume of experimental data (selective representations), while integration is performed over all values of arguments l; $A(q)$ and $A(l)$ are the corresponding weight functions.

The mean square discrimination functional of the second type has the form

$$\rho_\Theta(m) = \langle [\Theta(l|m) - \hat{\Theta}_q(l)]^2 \rangle, \qquad (7.3)$$

in which case probabilistic averaging (angular parentheses $\langle\rangle$) is defined in the sense that the estimate $\hat{\Theta}_q(l)$ is a random function, the value of which at a certain fixed l depends on the experiment number q.

These definitions of the functional $\rho_\Theta(m)$ differ primarily in that by using (7.2) we can find the actual value of the functional on the basis of experimental data, while by using (7.3) we can compute it, postulating some probabilistic properties of the estimate $\hat{\Theta}_q(l)$ as a random function.

The preference for one or the other of the functionals $\rho_\Theta(m)$ given above in conducting experiments is dictated chiefly by the specific nature of the problem to be solved. The selection is made upon examination of the more general system of which the system for conducting experiments is a subsystem. In some cases selection of the type of functional $\rho_\Theta(m)$ is made at a heuristic level on the basis of the conceptual content of the problem to be solved.

Expressions (7.2) and (7.3) can be discussed from the viewpoint of probability models and the measures of probability quantitatively describing these models. In fact, the characteristic $\Theta(l|m)$, which is contained in both of these expressions, is obtained in accordance with (6.3) on the basis of an *a priori* probabilistic description of the object of research being examined. In this sense, relationships (7.2) and (7.3), which are discrimination functionals for probability characteristics and statistical estimates, have at their basis the same probability model reflected in the characteristic $\Theta(l|m)$.

§8. The Base Model

It is natural to assume that values of functional $\rho_\Theta(m)$ will differ for different models $m \in M$, in view of the fact that hypotheses on the properties of the object of research reflect its true properties with different degrees of adequacy. Let us assume that among the set of models M there exists one for which functional $\rho_\Theta(m)$ is minimum.

We define the base model as $m_0 \in M$ which corresponds to a minimum value of functional $\rho_\Theta(m)$ -- that is,

$$\rho_\Theta(m_0) = \inf_{m \in M} \rho_\Theta(m), \tag{8.1}$$

where the symbol inf designates determination of the precise lower limit.

Obviously the closer the value of $\rho_\Theta(m)$ is to $\rho_\Theta(m_0)$ for a particular model m, the more correct and effective this model will be. The quantitative measure of effectiveness $E(m)$ of a certain model m can be defined as follows:

$$E(m) = \frac{\rho_\Theta(m_0)}{\rho_\Theta(m)}. \tag{8.2}$$

Effectiveness $E(m)$ of model m has the following properties:

$$E(m) < 1, \; m \neq m_0; \; E(m_0) = \sup_{m \in M} E(m) = 1, \tag{8.3}$$

where the symbol sup designates determination of the precise upper limit.

In forming the set of models M we must take account of all formalized *a priori* information a available and, in addition, we must consdier hypotheses h relative to the possible properties of the object studied. Let us turn our attention to some of the features behind forming set M of models m. In the general case, any two models m_i and m_j can be distinguished by volumes of formalized *a priori* information a_i and a_j being considered and by the hypotheses h_i and h_j assumed in defining these models. For the overwhelming majority of cases having practical interest, we can assume that the volumes of *a priori* information used to develop different models are equal, such that $a_i = a_j = a$, and that models $m_i = (a, h_i)$ and $m_j = (a, h_j)$ differ only with respect to the nature of the hypotheses h_i and h_j assumed. But even if this is not so, strictly speaking, the differences in volumes of *a priori* information being considered can be related in principle to characteristics of the corresponding hypotheses. Thus, under this interpretation differences in models stem from differences in the hypotheses assumed. If we designate by H a set of possible hypotheses $h \in H$, then, obviously, set M is defined uniquely by set H.

§9. Interpretation of Experimental Results

An important stage of experimental research is the interpretation of the results, which is defined as a procedure for making decisions on the properties of the real objects with the purpose of controlling subsequent experiments and developing recommendations for the use of the data obtained.

Inasmuch as we are examining quantitative experiments, we will discuss quantitative interpretations below.

We designate by $\Delta\rho_\Theta(m)$ the value of the functional $\rho_\Theta(m)$ corresponding to the event that the adopted model m matches the model of the real object precisely. It stands to reason that such a situation is hypothetical, but nevertheless it deserves examination from a methodological point of view. Considering the above discussion, we observe that values of $\Delta\rho_\Theta(m)$ should be treated as algorithmic and equipment errors in statistical measurement (this will be discussed in greater detail in Chapter VIII).

The value of $\Delta\rho_\Theta(m)$ can be computed if we assume the probability characteristic $\Theta(l|m)$ corresponding to the mth model to be the true characteristic of the object of research. We would naturally expect that $\Delta\rho_\Theta(m)$ would be minimum for the base model m_0 -- that is,

$$\inf_{m \in M} \Delta\rho_\Theta(m) = \Delta\rho_\Theta(m_0), \quad \Delta\rho_\Theta(m_0) \neq 0. \tag{9.1}$$

We can substantiate the validity of equation (9.1) in the following way. Statistical measurement errors (the algorithmic errors, at least) depend on the degree of correspondence of the adopted model to the real object, in which case the greater this correspondence is -- that is, the closer the model m being examined is to the base model m_0, the smaller these errors are (in fact, selection of the parameters of the statistical measuring system -- the smoothing time interval, the number of averaged selective executions, the resolution, and others -- is dependent upon the process adopted by the model). Following the definition of the base model, (8.1), it is closest to the true model among all $m \in M$. Thus, the validity of (9.1) is demonstrated.

Let ε be a positive number. Then the value

$$\Delta\rho_\Theta (m_0, \varepsilon) = (1 + \varepsilon) \Delta\rho_\Theta (m_0) \qquad (9.2)$$

corresponds to the minimum statistical measurement error multiplied by $1 + \varepsilon$. Of course, selection of a particular ε value must be justified, and the specific features of the problem to be solved and of using the experimental data must be taken into account. What the ε value does is provide a certain confidence range which determines the limits of permissible values for functional $\rho_\Theta(m)$.

If as a result of the experiment the condition

$$\rho_\Theta (m_0) < \Delta\rho_\Theta (m_0, \varepsilon) \qquad (9.3)$$

is satisfied, then the task of selecting the base model m_0 is fulfilled and the experiment completed. In the future, model m_0 can be used to predict the properties of the object of research under the set of conditions for which inequality (9.3) is defined.

But if we find that

$$\rho_\Theta (m_0) > \Delta\rho_\Theta (m_0, \varepsilon), \qquad (9.4)$$

then we cannot consider that the base model obtained adquately represents the real object (for the assumed value of ε, of course). In this case the set of possible models M should be replaced by another M_1 such that

$$M \cap M_1 = \varnothing, \qquad (9.5)$$

where the symbol $\cap$ designates an intersection of the sets and $\varnothing$ designates an empty set.

Further research reduces to searching for base model m_{10} among models $m_1 \in M_1$. If we find that

$$\rho_\Theta (m_{10}) < \Delta\rho_\Theta (m_{10}, \varepsilon), \quad (9.6)$$

then the experiment ends. But if an inequality reverse to (9.6) exists, then we must next introduce sets of models M_R, $R = 2,3,\ldots$, such that

$$\left(\bigcup_{j=1}^{R-1} M_j\right) \cap M_R = \varnothing, \quad (9.7)$$

and seek base model m_{R0} until the inequality

$$\rho_\Theta (m_{R0}) < \Delta\rho_\Theta (m_{R0}, \varepsilon) \quad (9.8)$$

holds for some m_{R0}

Next, let us examine some of the features of forming sets M_R, $R \geqslant 1$, of probability models. We recall that *a priori* information a and hypotheses $h \in H$ were used as initial data in the discussion of the set of models M.

If in view of (9.4) the necessity arises for shifting to model sets M_R, then they can be formed, on one hand, by adopting new hypotheses H_R, and on the other hand, with additional *a priori* information $a + \Delta a_R$, in which case the information Δa_R is obtained by conducting subsequent cycles of experimental research.

Another important consideration in this problem of interpretation is whether or not additional research must be conducted when we shift to sets M_R. If $\Delta a_R = 0$ -- that is, we do not receive new *a priori* information and, consequently, we form sets M_R on the basis of new hypotheses H_R, then, as a rule, the need for additional experiments does not arise. In this case the base models m_{R0} should be determined by measuring the characteristics of the same selective process that had been obtained previously. But if $\Delta a_R \neq 0$ and the volume of *a priori* information for forming models in M_R changes, then, as a rule, the conditions under which the experiments are performed change as well. This generates a need for performing additional experiments.

§10. Objects of Experimental Research in Sonar

Objects of experimental research in sonar are understood to be the water medium, its boundaries, and its irregularities (both of natural and aritifical origin), about the properties of which we can obtain information by performing statistical measurements on the characterisitcs of useful signals. Sonar methods are a most effective means for studying the underwater situation, since they enable us to change the type of signal emitted, the characteristics of the radiating and receiving antennas, and the algorithms for processing incoming information depending on the problem to be solved.

Thus, objects of research in sonar include the water medium,* the surface, the bottom, temperature irregularities, air bubbles, microorganisms, fish and other large marine animals, and artificial objects.

Let us now examine the characteristics of objects of research about which we could in principle gain information with sonar methods. Such characteristics are divided into three basic types -- spatial, temporal, and structural.

(i) Spatial characteristics include the distribution of objects in the water medium and the geometric characteristics of reflecting and scattering surfaces. Examples of such characteristics are the distribution of displacements on the water surface, the density of scattering units in space, roughness of the bottom, etc.

(ii) Temporal characteristics include the movement characteristics of objects. Examples of such characteristics are the distributions of the rates of movement of reflectors and scatterers, wave dynamics on the water surface, etc.

(iii) Structural characteristics include the characteristics of the materials composing the objects. Examples of such characteristics are the types of scatterers, bottom composition, etc.

---

*The refracting (focusing and defocusing) effects of the medium ($\nabla c \neq 0$) are not treated in this work; however, see [39]. (D.M.)

In more complex cases we can refer to spatial-temporal, spatial-structural, and structural-temporal characteristics of objects.

A key problem in using sonar research methods is that of determining the connection between the characteristics of objects and the characteristics of useful hydroacoustic signals. Equations relating the two types of characteristics are formulated in conjunction with the development of models of sonar signals. As a matter of course, one of the principal tasks of experimental research is testing the adequacy with which the models developed represent real objects -- that is, establishing an association between these equations and real dependencies and quantitatively evaluating this association.

Let us enumerate some of the probability characteristics of useful signals. The primary characteristics of this type are unidimensional probability distributions, autocorrelation functions, frequency spectra, and cross-correlation functions of signals observed at different points in space, cross-correlation functions of emitted and useful signals, and cross-spectra.

Of course, the list of informative signal characteristics is not exhausted by the ones mentioned here. We can also include two-dimensional probability distributions and their parameters, the characteristics of the envelope and the instantaneous phases of signals, and so on. The exploration of such characteristics, coupled with development of methods to measure them, is a pressing problem that has not been studied much as yet.

CHAPTER III

PROBABILITY MODELS OF HYDROACOUSTIC SIGNALS

§11. Varieties of Models

As had been noted above already, probability models developed through studies on the dynamic properties of objects of research are used to describe signals and interference in sonar (as we know, the dynamic properties of objects are described by their equations of motion with a consideration for limiting conditions). Today we can trace two approaches to the development of such dynamic models -- wave and phenomenological (see, in particular, [2,4,20-22, 31,35,39,42-44,49,60,63-65]). We will examine these approaches.

In the wave approach [4,31], the probability model is a solution to a wave equation, for which information on the characteristics of the rate of propagation of acoustic waves in the water medium, limiting conditions, and source characteristics are given as the initial data. We will refer to such probability models as wave models. In principal, wave models enable us to find direct relationships associating physical characteristics of the water medium, its boundaries, and the sources on one hand and various probability characteristics of useful hydroacoustic signals on the other. Wave models are highly substantiated with respect to physics. However, in developing such models we usually come to face the problem that it is impossible to conduct an adequately full probability analysis on signal properties even when the initial data are relatively simple (characteristics of the water medium, boundaries, and sources). Moreover, the initial data used to develop wave models are always idealized, and the models are described approximately and simplistically.

The second approach -- phenomenological -- is characterized by unrigid consideration of initial data and reduces to developing models that are simplified in the physical respect.* In this case, in sacrificing physical precision in formulating the problem during development of the model we rely on rather well known signal characteristics that can be measured easily. As a matter of fact, the procedure for developing the model reduces to determining an operator used in searching for concepts to describe hydroacoustic signals from among processes of simple types -- elementary processes. In this case constructing a probability model involves computing the probability characteristics of signals on the basis of corresponding characteristics of elementary processes. In most cases construction of phenomenological models is based on the physical intuition of the investigator, on his experience, and is conducted at a heuristic level.

At least three types of phenomenological models are possible -- *canonical*, *constructive*, and *parametric*.

Canonical (received) signal models are based on representing signals in the form of a sum of elementary oscillations having random parameters or in the form of an integral transformation of elementary oscillations, the characteristics of which are random. The former are referred to as discrete canonical models and the latter as integral canonical models. Therefore, in this case we define canonical concepts (cf. V. S. Pugachev [25]) of signals as random functions. Here we assume that the probability characteristics of the random parameters of the elementary oscillations are known, and then the task of analyzing signals reduces to the following: Using known probability characteristics of the random processes of elementary oscillations we must determine probability characteristics of hydroacoustic signals which are significant to the problem being studied.

There are definite differences between canonical concepts for random functions (cf. V. S. Pugachev) and the canonical models of hydroacoustic signals under consideration here. With respect to canonical concepts the task is

---

*This includes simplified boundary conditions and propagation mechanisms (wave equations, etc.), for example. (D.M.)

reduced to finding a method for expressing the random function of interest by discrete (pulsed) or continuous white noise and some nonrandom, so-called coordinate functions. Such a procedure bears a considerably formal mathematical nature. In developing canonical models of hydroacoustic signals we begin with the fact that the properties of the elementary oscillations are known from the physical conditions of the problem of interest and, in addition, that the coordinate functions can themselves depend on random parameters. Nevertheless, there is an extremely important similarity between canonical concepts describing random functions and canonical models of hydroacoustic signals: Both reflect a tendency to express random processes by simpler functions. This considerably simplifies the solution of many probability problems. Moreover, it appears to us that canonical concepts describing random functions have a deep physical meaning, which is revealed to a certain extent through the development of canonical probability models of signals. The development of such models in sonar is treated in [14-16,19,22,33,35,36,39,52].

Constructive models of hydroacoustic signals are based on representing the latter in the form of combinations (additive-multiplicative in the general case) of some elementary random processes. In this case we assume a certain algorithm for forming the random process of interest to us (the signal) out of elementary processes and consider that the probability characteristics of these elementary processes are known. Then, analysis of hydroacoustic signals is reduced to the following: Using known probability characteristics of elementary random processes we must determine the probability characteristics of the hydroacoustic signals which are significant to the problem being solved. The development of constructive models in sonar is treated in [2,21,22,34,53,56,57,62].

Parametric models of signals are based on representing the latter in the form of deterministic time functions depending on random parameters. Such models are extremely simple and are adequately effective in a number of cases. In some sense they can be looked upon as particular cases of constructive models.

As follows from the discussion above, probability models assume the following order with respect to the completeness with which physical properties of objects of research are taken into account: Wave models take the fullest account, then canonical, and finally constructive and parametrical. But if we examine the completeness of the probability analysis performed on the signals, the order of the models reverses: Parametrical and constructive models provide the most complete probability analysis, then canonical, and finally wave.

We emphasize that initial (*a priori*, generally) information is used differently in the various probability models described. Initial information is considered most fully in wave models, but it is difficult to use in view of the difficulty in solving wave equations.* This information is taken into account less fully in canonical models, but it can be used more extensively than in the previous case. Finally, initial information is used least fully in constructive and parametrical models, but it can be used most extensively. Of course, it would be desirable to develop combined models as well. In particular it is extremely important to incorporate methods of ray acoustics to represent hydroacoustic fields as ray tubes, linear congruences, and so on when developing phenomenological models of hydroacoustic signals. In this case we can achieve both effective consideration of the physical properties of the object of research and an adequately complete probability analysis of the signals.

We should note one more aspect of importance in selecting the means for constructing probability models of signals and interference in sonar. From the point of view of so-called systemic problems (which statistical problems of sonar are), it would be desirable to describe the water medium, objects of research, and sources in the language of equivalent quadripoles,† the characteristics of which associate input variables to output effects in a random fashion. This approach has been employed extensively in [1,12-14,22, 25,34-36,37,51,53,63]. In terms of our symbols (see §2) we must find, in particular, the phenomenological equivalents of operators M_1, m_1, M_2, m_2, T, and s in the form of

---

*However, see [63]. (D.M.)

†Equivalent tripole networks with two input and two output terminals; e.g. "filters." (D.M.)

certain characteristics (transient functions, frequency characteristics) of quadripoles with random properties. On one hand, such equivalents must be realistic enough from the viewpoint of their closeness to the objects of research and, on the other, they must be easy to use in systemic problems.

This discussion leads us to the conclusion that it would be suitable to make wide use of phenomenological models in sonar.

§12. *A Priori* Information and Probability Models

In examining the structural diagram for conducting sonar research we had noted that probability models of signals are developed with a consideration for the *a priori* information available and the particular hypotheses assumed. Below we will discuss some problems involved with the means for taking account of *a priori* information when developing probability models of hydroacoustic signals.

A priori information may be diverse. Three basic types are distinguished -- information on properties of the object of research, on the class of hydroacoustic signal, and on the probability characteristics of the hydroacoustic signal.

Use of *a priori* information is illustrated by the diagram in Figure 3. We note that we must determine four characteristics, namely the operators m and μ_m as well as the probability characteristics $\Theta(l|m)$ and $\pi_i(\lambda)$.

A priori information on the dynamic properties of the object of research affects selection of the model operator m and the form of the probability characteristics $\pi_i(\lambda)$ of initial random processes $\xi_i(t)$ out of which hydroacoustic signals, $x(t)$ are formed. Information on the class of hydroacoustic signal defines the model operator m as well as the type of characteristic $\Theta(l|m)$, which corresponds to the class of the signal. Finally, information on probability characteristics of the hydroacoustic signal is taken into account in determining $\Theta(l|m)$. We note that operator μ_m is defined uniquely by the type of operator m for the particular model selected.

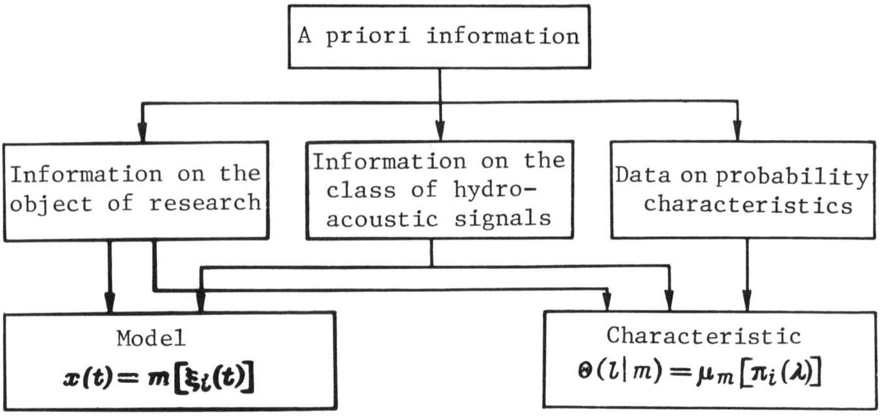

Figure 3. *A priori* information use diagram.

The means for taking account of formalized information is extremely important. It would be desirable for this method to be adequately simple and to take account of *a priori* information to the fullest extent possible. An analysis of this process demonstrates (see, for example, [12-16,22,35,39]) that phenomenological models of random processes are the best in this regard. By representing signals as random processes at the output of equivalent networks, we can construct many adequate models of hydroacoustic information [63].

§13. Methods for Describing Hydroacoustic Signals in Terms of Probability

Let us examine a random process $x(t)$ corresponding to the hydroacoustic signal model. Let $x_k(t)$ be the kth representation of process $x(t)$ and $\{x_k(t)\}$ be the set of these representations.

There exist three methods for defining the probability characteristics of random process $x(t)$, each of which corresponds to a certain method for describing the process [20,23,24]. Prior to ecamining these methods we will note some typical traits of the concept "random process":

(a) The concept "random process" has hypothetical experiments at its basis, through which we obtain a set of random time functions;

(b) A random process is an idealized reflection of a set of random functions coupled with a given measure of probability (probability distributions);

(c) Random process $x(t)$ can be viewed either as a set of representations $x_k(t)$ depending on parameter k, or as a random variable depending on time t for a fixed representation number (or index) k.

Next, we will discuss methods for describing random processes existing as models of hydroacoustic signals in terms of probability, with a consideration for the points formulated above.

Let ϑ be an operator for transformation of $x_k(t)$ that corresponds to a probability characteristic of type Θ. The first method for describing random process $x(t)$ in terms of probability reduces to an examination of a set of representations $\{x_k(t)\}$ at a fixed point of time t (or at several fixed times $t_1, t_2,\ldots,t_n$). In this case random process $x(t)$ is treated as a random variable (or as a set of variables in the more general case), the value of which changes depending on the representation number k. Consequently, the probability characterisitc $\Theta(t,l)$ is described in the following way:

$$\Theta(t,\ l) = \langle \vartheta\,[x_k(t),\ l] \rangle = \lim_{N\to\infty} \frac{1}{N} \sum_{k=1}^{N} \vartheta\,[x_k(t),\ l]. \qquad (13.1)$$

Such a characteristic reflects the possible dependence of the probabilistic properties of the process on time t and is called the current-t (i.e., instantaneous-t) probability characteristic (we note that the angle brackets in (13.1) represent obtaining characteristic $\Theta(t,l)$ averaged over the ensemble of representations of the random process). When we describe the random process in this way, developing its model should reduce to formulating representations of the given process $\{x_k(t)\}$ at a fixed point in time (or at several fixed times) by some random variables or functions. As follows from (13.1), we must examine the ensemble of the process's representations in determining the current-t probability characteristic so that averaging over the ensemble can be carried out.

The second method for describing a random process in terms of probability reduces to examining a particular fixed representation $x_k(t)$ of the process. In this case random process $x(t)$ is treated as a random function $x_k(t)$, the values of which change depending on time. The probability characteristic $\Theta(k,l)$ takes the following form:

$$\Theta(k, l) = \overline{\vartheta[x_k(t), l]} = \lim_{T\to\infty} \frac{1}{T} \int_{-T/2}^{T/2} \vartheta[x_k(t), l]\, dt. \qquad (13.2)$$

Such a characteristic reflects the possible dependence of the probabilistic properties of the process on the number k of its representation and is called the current-k probability characteristic (we note that the expression in (13.2) covered by the continuous line represents averaging characteristics $\Theta(k,l)$ over time). Such a method for describing a random process requires that we develop a model of it such that each representation $x_k(t)$ is designated by certain initial random variables or functions. As follows from (13.2), in determining the current-k probability characteristic we must examine each representation of the process so that the representation can be averaged over time.

The third and final method for describing a random process in terms of probability reduces to examining a set of representations $\{x_k(t)\}$ in coordinates (t,k) simultaneously. In this case the probability characteristic $\Theta(l)$ takes the following form:

$$\Theta(l) = \overline{\langle\vartheta[x_k(t), l]\rangle} = \lim_{T\to\infty} \lim_{N\to\infty} \frac{1}{NT} \sum_{k=1}^{N} \int_{-T/2}^{T/2} \vartheta[x_k(t), l]\, dt. \qquad (13.3)$$

Such a characteristic defines the probabilistic properties of the random process as a whole and is called the average probability characteristic.

The following two relationships are obvious:

$$\Theta(l) = \overline{\Theta(t, l)}, \quad \Theta(l) = \langle\Theta(k, l)\rangle. \qquad (13.4)$$

Thus, the current-t probability characteristic $\Theta(t,l)$ gives us the functional dependence of the process's properties on time, the current-k probability characteristic $\Theta(k,l)$ gives us the functional dependence of the process's properties on the number of the representation, while the average probability characteristic defines the probabilistic properties of the process as a whole.

The examination of the varieties of probability characteristics of random processes, namely the current-t $\Theta(t,l)$, the current-k $\Theta(k,l)$, and the average characteristics $\Theta(l)$, sheds light on the problem of obtaining the ensemble of the process's representations $\{x_k(t)\}$. We will discuss this problem in more detail.

One kth experiment (trial) conducted with respect to time and having random results with respect to time is the experimental (hypothetical) analog of each kth representations of $x_k(t)$. The ensemble of such experiments conducted simultaneously produces the ensemble of representations $\{x_k(t)\}$. In a number of cases we can ignore the need for conducting the experiments simultaneously. Such a situation is possible when the investigator possesses information for points in time corresponding to identical (in the static sense) experimental conditions [53,58,62]. In this case we can construct an ensemble of representations without resorting to parallel experiments, but rather by ordering time functions and transforming them into representations. Sonar provides a typical example of constructing an ensemble of representations. In this case the points in time mentioned above are the moments at which signals are emitted into the water medium such that, for example, examination of the ensemble of echo signals or reverberatory signals could be reduced to placing them in order with the times of signal emission matching.

If such points in time, which we use to place representations of the process in an order, cannot be determined, then it is impossible to obtain an ensemble of representations by conducting experiments based on time. The only possible way to construct an ensemble of process representations in this case (in the absence of a possibility for conducting experiments that are parallel in time) would be to construct hypothetical models of the representations. In this event the types of initial data -- *a priori* information and the hypotheses assumed -- must be distinguished clearly.

§14. Classification of Hydroacoustic Signals

We can classify hydroacoustic signals generally on the basis of the current-t $\Theta(t,l)$, current-k $\Theta(k,l)$, and average $\Theta(l)$ probability characteristics of the signals (as random processes). We will consider the dependence (or independence) of current characteristics on time (current-t characteristic) and on the number of the representation (current-k characteristic) to be the main index of such a classification.

If the equality

$$\Theta(t, l) = \Theta(l) \tag{14.1}$$

holds, then the random process is stationary. In this case, as follows from (14.1), the current-t probability characteristic is not a function of the point in time at which reading begin and is the same as the average characteristic. In the opposite case, when condition (14.1) does not hold, the process is nonstationary. We can distinguish at least two types among nonstationary random processes:

1. Processes which are stationary over an interval for which the following condition holds:

$$\Theta(t, l) = \Theta_i(l), \quad t \in [t_i - T_i/2,\ t_i + T_i/2]. \tag{14.2}$$

2. Periodically nonstationary processes having the following equality as to their index:

$$\Theta(t, l) = \Theta(t + jT_0, l), \quad j = 0, \pm 1, \ldots \tag{14.3}$$

If the relationship

$$\Theta(k, l) = \Theta(l) \tag{14.4}$$

holds, then the process is (statistically) homogeneous. In this case, according to (14.4) the current-k probability characteristic is not a function of the number of the representation selected and is the same as the average characteristic. In the opposite case, when condition (14.4) does not hold, the process is heterogeneous (statistically inhomogeneous). As in the case of nonstationary processes, we can distinguish at least two types among the heterogeneous processes:

1. Homogeneous processes with a finite ensemble of representations for which the following condition holds:

$$\Theta(k, l) = \Theta_i(l), \ k \in [k_i - n_i/2,\ k_i + n_i/2]. \tag{14.5}$$

2. Periodically heterogeneous processes with the following equality as to their index:

$$\Theta(k, l) = \Theta(k + jn_0, l), \qquad j = 0, \pm 1, \ldots \qquad (14.6)$$

Finally, if

$$\Theta(t, l) = \Theta(k, l) = \Theta(l), \qquad (14.7)$$

then we have an ergodic random process. In the event that (14.7) does not hold, the process is nonergodic and can be treated as one of the types of nonstationary or heterogeneous processes. In this case neither the current-t nor the current-k probability characteristic is the same as the average characteristic.

A classification of random processes corresponding to hydroacoustic signals of the types examined above is shown in Figure 4. This is an enlarged classification, but nevertheless it does provide important information to the investigator as he develops probability models and selects the type of probability characteristic to be studied.

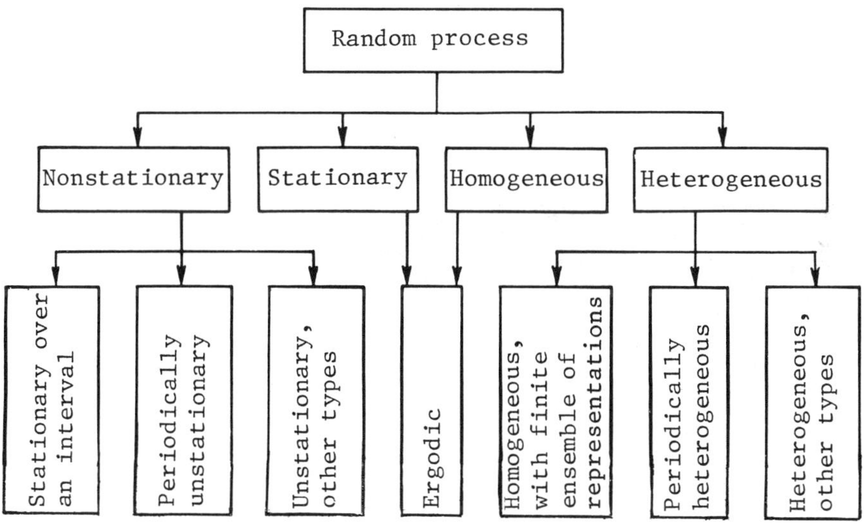

Figure 4. Classification of hydroacoustic signals as random processes.

Selection of a particular type of characteristic -- $\Theta(t, l)$, $\Theta(k, l)$, or $\Theta(l)$ -- is governed by the problems posed in conducting research on properties of hydroacoustic signals. For example, if it is known that these signals are nonstationary, then the current-t characteristic should be studied, inasmuch as only it can represent the characteristics of nonstationary signals. To develop probability models of hydroacoustic signals on this basis we must have, first, information on the probability characteristics of random variables defining the behavior of the objects of research within the set of hypothetical conditions, and, secondly, data on the nature of nonstationary signals. If it is known that hydroacoustic signals are heterogeneous, then we should study the current-k characteristic, inasmuch as only it can represent the nature of heterogeneous signals. To develop probability models of such hydroacoustic signals we must have, first, information on the probability characteristics of random functions defining the behavior of the objects of research with respect to time and, secondly, data on the nature of heterogeneity of the signals.

It does not matter which of the probability characteristics enumerated above is studied in regard to stationary and homogeneous -- that is, ergodic -- signals, since all information on the process's properties is contained in either of the representations. However, models of this type are so idealized that their use is improper in most cases of practical interest.

We note that the stationarity (or nonstationarity) and homogeneity (or heterogeneity) of the signals can often be defined only as hypotheses, and not as *a priori* information. In this case the experiments should involve testing these hypotheses, for which it would be necessary to compare models of hydroacoustic signals $x(t)$ with selective processes $\hat{x}(t)$ obtained in the course of experimental research. The procedures of such a comparison had been examined in §6.*

§15. Canonical Models

The canonical representations of random functions developed by V. S. Pugachev [25] enable us to construct

*(See also, especially, [53].) (D.M.)

canonical models of several types of hydroacoustic signals. We will examine such models and interpret them below.

As was noted in §11, canonical signal models can be separated into discrete and integral models. Let $x(t)$ be a random process describing the signal. Then a concept of the form

$$x(t) = \sum_{k=1}^{n} V_k \varphi_k(t) \tag{15.1}$$

is a discrete canonical model of the signal, where V_k are uncorrelated (independent in the more general case) random functions and $\varphi_k(t)$ are nonrandom time functions. The number of terms n in sum (15.1) can be finite or infinite (note that $\langle x(t) \rangle = 0$).

A model such as (15.1) can describe hydroacoustic signals arising through the scattering of waves off of discrete irregularities in the water medium and off of its boundaries, as well as through multiple-beam propagation of the signals. Such models are examined in [1,2,34] applicable to multiple-beam propagation of signals and in [13-16,19,33,35] in regard to studies on the probability characteristics of reverberatory signals. It should be noted that the effectiveness of canonical models has been supported many times in experimental research on the properties of real hydroacoustic signals.

The following is a somewhat more general form of the discrete canonical model:

$$x(t) = \sum_{k=1}^{n} V_k \Psi_k(t, \varepsilon_k), \tag{15.2}$$

where ε_k is a set of random parameters*, $\Psi_k(t, \varepsilon_k)$ is a nonrandom time function that can have, in particular, the form

$$\Psi_k(t, \varepsilon_k) = \int_{-\infty}^{\infty} h_k(t - t', \varepsilon_k)\, \varphi_k(t')\, dt', \tag{15.3}$$

* ε_k can be, for example, source coordinates, points of observation, and other parameters of the objects of research.

where $h_k(t, \varepsilon_k)$ is a certain transient function characterizing a linear transformation of $\varphi_k(t)$. The function $h_k(t, \varepsilon_k)$ can be interpreted physically when, for example, constructing a scattering model, as: If $K_k(\omega, \varepsilon_k)$ is a complex frequency characteristic of the kth scatterer, then

$$h_k(t, \varepsilon_k) = \frac{1}{2\pi} \int_{-\infty}^{\infty} K_k(\omega, \varepsilon_k) \exp(j\omega t)\, d\omega. \tag{15.4}$$

In the case of multiple-beam hydroacoustic signals the function $h_k(t, \varepsilon_k)$ may describe absorption corresponding to the kth path of the signal's propagation.

Next, let us examine the integral canonical model of signal $x(t)$. This model involves the following signal representation:

$$x(t) = \int_{-\infty}^{\infty} V(t')\, \varphi(t', t)\, dt', \tag{15.5}$$

where $V(t')$ is a random function having the properties of white noise and $\varphi(t', t)$ is a nonrandom function.

A model such as (15.5) can describe hydroacoustic signals arising as a result of the reflection of waves from various types of objects of detection, as well as signals scattered off of various types of irregularities. The following model of the signal may be more general than (15.5):

$$x(t) = \int_{-\infty}^{\infty} V(t')\, \Psi(t', t, \varepsilon)\, dt', \tag{15.6}$$

where ε is a set of random parameters on which the form of the function $\Psi(t', t, \varepsilon)$ depends.

Canonical models of hydroacoustic signals yield most easily to probability analysis and correspond to the physical aspects of signal formation with adequate fullness. It should be noted that such models are most useful when the hydroacoustic signals being studied are made up of a set of elementary signals with random parameters, and when the signals could be conceived of as the result of passage of an elementary signal through a certain equivalent system (generalized filter) with random parameters (or characteristics).

§16. Parametrical Models

A parametrical model of signal $x(t)$ is such that the signal is represented by:

$$x(t) = \xi(t, \alpha), \qquad (16.1)$$

where ξ is a certain known function that is nonlinear (in t) in the general case, and $\alpha = (\alpha_1, \alpha_2, \ldots, \alpha_n)$ is a set of independent random variables.

The parametrical model (16.1) is a constructive signal model, the simplest of that type. Such models can be used to describe echo signals in sonar when their differences from emitted signals are expressed in parametrical form. In some cases such models correspond to signals observed at the output of sonar information processing systems. Parametrical models are used rather extensively in [1,2,22,36] to describe signals propagating in heterogeneous media, and echo signals rebounding from objects of detection.

§17. Additive-Multiplicative Constructive Models

A rather general constructive model of the hydroacoustic signal $x(t)$ is achieved when the signal is represented in the form of an additive-multiplicative combination of processes $\xi_i(t)$, $i = 1,2,3,4$:

$$x(t) = \xi_1(t)\,\xi_2(t) + \xi_3(t)\,\xi_4(t). \qquad (17.1)$$

Here either all $\xi_i(t)$ can be random processes, or some of them can be random processes while the rest are known functions. Expression (17.1) embraces models of practical interest describing random amplitude, phase, and frequency modulation of signals of a known type (emitted signals in sonar), and models describing the effect of an additive random component on the signal. Models such as (17.1) give a rather good description of hydroacoustic signals propagating in a heterogeneous water medium, as well as of echo signals formed when waves are scattered and when objects of detection move. Constructive models of hydroacoustic signals such as (17.1) are used in a number of papers [2,20-22] to perform probability analyses on echo signals scattered from various types of objects of research.

§18. Models in the Form of Oscillations with Random Amplitude and Phase

In this case signal $x(t)$ is represented by

$$x(t) = E(t) \cos \Phi(t), \tag{18.1}$$

where $E(t)$ is the envelope and $\Phi(t)$ is the current (i.e., instantaneous) phase of the signal. Obviously, model (18.1) is a particular case of (17.1) in which $\xi_1(t) = E(t)$, $\xi_2(t) = \cos \Phi(t)$, $\xi_3(t) = \xi_4(t) = 0$.

If the signal has a certain centrally occupied frequency or, as it is sometimes called, "carrier frequency," equal to ω_0, then we can assume that

$$\Phi(t) = \omega_0 t + \varphi(t). \tag{18.2}$$

Then

$$x(t) = E(t) \cos [\omega_0 t + \varphi(t)], \tag{18.3}$$

in which case $E(t)$ and $\varphi(t)$ are, as a rule, slowly changing functions as compared to $\cos \omega_0 t$, $\sin \omega_0 t$.

It should be kept in mind that if no additional conditions are imposed, functions $E(t)$ and $\varphi(t)$ in (18.3) cannot be defined uniquely. In fact, we can use the most diverse combinations of $E(t)$ and $\varphi(t)$ to obtain $x(t)$. However, we are interested in cases in which these functions have a completely specific physical meaning and their selection cannot be arbitrary. This question will be discussed in the next paragraph.

§19. Complex Representations of Signals*

The signals $x(t)$ that we had examined above are real functions of time. But in regard to many problems we find it convenient to introduce complex representations of signals and examine the so-called "analytical signal" $Z(t)$, which is related to the initial signal $x(t)$ by the following relationship:

---

*[12,46,47,51].

$$Z(t) = x(t) + j\tilde{x}(t), \tag{19.1}$$

where

$$\tilde{x}(t) = -\frac{1}{\pi}\int_{-\infty}^{\infty}\frac{x(t')\,dt}{t-t'} \tag{19.2}$$

is the Hilbert transformation of $x(t)$ such that*

$$x(t) = \frac{1}{\pi}\int_{-\infty}^{\infty}\frac{\tilde{x}(t')\,dt}{t-t'}. \tag{19.3}$$

We use a complex (analytical) signal $Z(t)$ primarily to simplify computations, but in reality we deal with the real signal when conducting research. Complex representations of signals are used extensively in performing probability analyses on hydroacoustic information, particularly on emitted signals, marine reverberation, and echo signals. We note that

$$x(t) = \operatorname{Re} Z(t), \quad \tilde{x}(t) = \operatorname{Im} Z(t). \tag{19.4}$$

As with any sort of complex function, $Z(t)$ can be presented in exponential form:

$$Z(t) = A(t)\exp[j\Phi(t)], \tag{19.5}$$

or, equivalently,

$$Z(t) = A(t)\cos\Phi(t) + jA(t)\sin\Phi(t). \tag{19.6}$$

Relationships (19.1) and (19.6) provide unique definitions of envelope $A(t)$ and the instantaneous phase $\Phi(t)$ of the signal $x(t)$:

$$A(t) = [x^2(t) + \tilde{x}^2(t)]^{1/2}; \tag{19.7}$$

$$\Phi(t) = \arctan\frac{\tilde{x}(t)}{x(t)}. \tag{19.8}$$

For quasiharmonic (i.e., narrow band) signals, on the basis of (18.2) and (19.5) we have:

---

*We have in mind the principal values of the integrals in expressions (19.2) and (19.3) in the Cauchy sense.

$$Z(t) = A(t) \exp[j\omega_0 t + j\varphi(t)] = Z_x(t) \exp(j\omega_0 t), \qquad (19.9)$$

where the function

$$Z_x(t) = A(t) \exp[j\varphi(t)] \qquad (19.10)$$

is called the *complex envelope* of the signal and defines its joint amplitude-phase modulation.

CHAPTER IV

THE EMITTED SIGNALS

§20. Signal Classification

One of the important features of sonar as a method for studying underwater objects is the possibility for using different types of emitted signals.

Conflicting requirements are imposed on the emitted signal in a number of cases. In fact, on one hand it is desirable to decrease the length of the signal in order to, for example, reduce the effects of reverberatory interference, increase spatial resolution, and so on, while on the other hand it would be suitable to increase this length so as to decrease the effects of noise interference, attain higher precision in measuring the angular coordinates of the object of research, and so on. The situation is similar in regard to other parameters of the emitted signals. We can discuss selection of an efficient type of emitted signal in terms of the nature of the research problem and the experimental precision required.

The problem of selecting a particular type of emitted signal becomes more or less defined when we limit our consideration of signals to just one particular class, within which all of the varieties of the signal can be obtained by changing the parameters or form of modulating functions.

We will examine pulsed sonar systems. This approach imposes the following conditions on the time function $C(t)$ describing the emitted signal* during one emission-reception cycle:

$$\int_{-T_{e\text{-}r}/2}^{T_{e\text{-}r}/2} C^2(t)\,dt < T_{e-r} \sup_t C^2(t), \tag{20.1}$$

---

*See footnote, p. 109 (D.M.)

where T_{e-r} is the time interval between successively emitted signals. What condition (20.1) actually means is that the effective duration T_{ef} of the emitted signal, which depends upon its energy, is less than time interval T_{e-r} corresponding to one emission reception cycle -- that is,

$$T_{\text{ef}} < T_{e-r}.$$

If we ignore the "tails" of function $C(t)$, which are significant with respect to energy, to say that the emitted signal is pulsed means that its energy is finite for one emission-reception cycle:

$$\int_{-\infty}^{\infty} C^2(t)\,dt < \infty.$$

We distinguish between simple and complex signals among the emitted signals used in pulsed sonar.

Simple signals are defined as those for which the parameter

$$\mu = \Delta F_{\text{ef}} T_{\text{ef}}, \qquad (20.2)$$

which is the product of the effective spectrum width ΔF_{ef}, and duration T_{ef}, is constant and equal to $\mu \simeq 1$. Such signals include those with sinusoidal occupation and various forms of envelopes. It is typical for a simple signal that if the form of its envelope is given, then parameters ΔF_{ef} and T_{ef} are associated uniquely (this corresponds to the condition μ = const for signals of this form).

Complex signals are defined as those for which parameter (20.2) is much larger than one -- that is, $\mu > 1$. Absence of a unique association between parameters ΔF_{ef} and T_{ef} is typical of complex signals. In this case the effective spectrum width ΔF_{ef} of the signal depends not so much on the signal's effective duration T_{ef} as on the characteristics of the modulating functions, namely the deviation and rate of change of phase or frequency in the case of phase- or frequency-modulated signals, the characteristics of modulating random processes in the case of signals with random amplitude, phase, and frequency modulation and, finally, the spectrum width for initial random noise in the case of a signal representing an interval of a stationary random process.

Explosive signals and series of simple signals occupy an intermediate position between simple and complex signals. Explosive signals are typically wide-band signals for which the complexity factor μ can equal several units. Series of simple signals do not possess a smooth spectrum, and their ΔF_{ef} and T_{ef} parameters can only be defined conditionally. A certain equivalent complex signal can be compared to each series of signals.

We will examine the characteristics of most of the enumerated signals below.

§21. Complex Representation of Emitted Signals

Given sufficient general assumptions, the emitted signal $C(t)$ can be represented in the form

$$C(t) = A(t) \cos [\omega_0 t + \Phi(t)], \qquad (21.1)$$

where $A(t)$ is the envelope characterizing change of signal amplitude with respect to time, ω_0 is the central (carrier) frequency, and $\Phi(t)$ is the phase modulation law. We can also refer to the so-called instantaneous frequency of the signal

$$\omega_C(t) = \omega_0 + \frac{d}{dt}\varphi(t)$$

or to its instantaneous phase

$$\varphi(t) = \int_0^t \omega_C(t')\,dt'.$$

Function $C(t)$, which describes the emitted signal, is real and exists in real time. However, in solving many problems it is convenient to use a complex form to represent the emitted signal, similar to the way described in §19.

An emitted signal written in the complex form

$$Z(t) = C(t) - j\tilde{C}(t)$$

is called an analytical signal, in which case $\tilde{C}(t)$ is a component of the signal existing in a Hilbert conjugation with signal $C(t)$.

Taking (21.1) into account, we write $Z(t)$ in exponential form

$$Z(t) = A(t) \exp[-j\omega_0 t - j\Phi(t)] \tag{21.2}$$

and introduce the complex envelope

$$Z_C(t) = A(t) \exp[-j\Phi(t)], \tag{21.3}$$

characterizing the joint amplitude-phase law of modulation of the emitted signal. In this case it follows from (21.2) and (21.3) that

$$Z(t) = Z_C(t) \exp(-j\omega_0 t). \tag{21.4}$$

The reason we introduce the complex envelope of the signal $Z_C(t)$ is that we thus remove the unessential factor $\exp(-j\omega_0 t)$, which bears no information regarding the form of the initial signal.

We note that the spectrum of the analytical signal $Z(t)$ matches in form the spectrum of the real signal $C(t)$ (both spectra differ from each other by a scale factor 2). In addition, by shifting from signal $Z(t)$ to complex envelope $Z_C(t)$ we reduce the spectrum to null frequency without changing its form. Therefore, use of the complex envelope of the emitted signal, which is convenient from the standpoint of computation, does not make any changes in the method for describing the signals.

Various instrumental methods exist for obtaining the complex envelope of a signal. In particular, we find it convenient to use the so-called quadrature components $Z_{Cc}(t)$ and $Z_{Cs}(t)$ of the signal's complex envelope $Z_C(t)$. These components are obtained from the expression

$$Z_C(t) = Z_{Cc}(t) - jZ_{Cs}(t), \tag{21.5}$$

in which case it follows from (21.3) and (21.5) that

$$\left.\begin{aligned} Z_{Cc}(t) &= A(t) \cos\Phi(t), \\ Z_{Cs}(t) &= A(t) \sin\Phi(t). \end{aligned}\right\} \tag{21.6}$$

Various methods for obtaining the *quadrature components* of hydroacoustic signals are described in [14,22,35, 53,66].

§22. Ambiguity Function* and Diagrams

It is expedient to describe emitted signals by some single method so that they can be compared in terms of equivalent, sufficiently universal characteristics. We use the so-called ambiguity function for this purpose, introduced originally (in application to radar) by Woodward [6] and utilized extensively to analyze the characteristics of radar and sonar signals [6,14,22,32-37,46,50,51,67].

By definition, the ambiguity function $\chi(\Omega,\tau)$ is written in the form

$$\chi(\Omega, \tau) = \frac{1}{E_C} \int_{-\infty}^{\infty} Z_C(t-\tau) Z_C^*(t) \exp(j\Omega t)\, dt, \qquad (22.1)$$

where Ω and τ are, correspondingly, frequency and time shifts, $Z_C^*(t)$ is a function complexly conjugated with $Z_C(t)$, and

$$E_C = \int_{-\infty}^{\infty} |Z_C(t)|^2\, dt$$

is the energy of the signal's complex envelope $Z_C(t)$.

As can be seen from (22.1), the ambiguity function $\chi(\Omega,\tau)$ of the emitted signal characterizes the degree of match between signals, one of which is shifted in time by the amount τ -- the factor $Z_C(t - \tau)$, while the other, the complexly conjugated signal, is shifted in frequency by the amount Ω -- the factor $Z_C^*(t) \exp(j\Omega t)$. Thus, the ambiguity function provides a universal description (at the correlation level) of the emitted signal in the frequency-time domain. Of course, such a description has advantages over temporal and over frequency descriptions of signals. In addition, as we will see below, the ambiguity function defines a whole series of important characteristics of sonar systems.

It should be kept in mind that the ambiguity function of signals arose as a result of the fact that a cross-correlation system with a copy of the emitted signal shifted in time and frequency and an input process as its inputs

---

*Called "indeterminacy" function in the original. (D.M.)

was found to be optimum when signals of a known form are received on a background of Gaussian white noise [6,13,32, 36,37,39,47,51,67]. However, if the interference is not Gaussian or cannot be interpreted as white noise or, finally, if the useful signal does not match the emitted signal in form, a cross-correlation detection system would not be optimum in the strict sense of the word. Nevertheless, it can be used under these conditions in view of its simplicity, as well as for a number of other reasons.

Here are some of the properties of the ambiguity function [16,31,32,46,47,50,67]:

1) The volume of the "solid of ambiguity" is constant:

$$\int_{-\infty}^{\infty}\int_{-\infty}^{\infty} |\chi(\Omega, \tau)|^2 \, d\Omega \, d\tau = 2\pi; \tag{22.2}$$

2) the ambiguity function is expressed in terms of the spectral density

$$g_{Z_C}(\omega) = \int_{-\infty}^{\infty} Z_C(t) \exp(-j\omega t) \, dt$$

of the signal's complex envelope $Z_C(t)$:

$$\chi(\Omega, \tau) = \frac{1}{2\pi E_C} \int_{-\infty}^{\infty} g_{Z_C}(\omega) \overset{*}{g}_{Z_C}(\omega + \Omega) \exp(j\omega\tau) \, d\omega;$$

3) sectioning function $\chi(\Omega,\tau)$ in a plane perpendicular to (Ω,τ) and passing through line $\tau = 0$ produces

$$\chi(\Omega, 0) = \frac{1}{E_C} \int_{-\infty}^{\infty} |Z_C(t)|^2 \exp(-j\Omega t) \, dt$$

$$= \frac{1}{2\pi E_C} \int_{-\infty}^{\infty} g_{Z_C}(\omega) \overset{*}{g}_{Z_C}(\omega + \Omega) \, d\omega,$$

which corresponds to the frequency spectrum of the square of the modulus of the signal's complex envelope, or to the convolution of its complex spectra;

4) sectioning function $\chi(\Omega,\tau)$ in a plane perpendicular to (Ω,τ) and passing through line $\Omega = 0$ produces

$$\chi(0, \tau) = \frac{1}{E_C} \int_{-\infty}^{\infty} Z_C(t-\tau) Z_C^*(t)\, dt$$

$$= \frac{1}{2\pi E_C} \int_{-\infty}^{\infty} |g_{Z_C}(\omega)|^2 \exp(j\omega\tau)\, d\omega.$$

The function $\chi(0,\tau)$ corresponds to the so-called signal autocorrelation function or to a Fourier transform of the square of the spectrum modulus of the signal's complex envelope.

The form of the ambiguity function is usually defined by the square of its modulus $|\chi(\Omega,\tau)|^2$, and a so-called ambiguity diagram is used to gain an orientational assessment of the form of $|\chi(\Omega,\tau)|^2$. We note that, sometimes, instead of function $\chi(\Omega,\tau)$, the square of its modulus, $|\chi(\Omega,\tau)|^2$, is called the ambiguity function.

There are several methods for setting up the ambiguity diagram, but they are all based on selecting a fixed level k which, as follows from the condition

$$|\chi(\Omega, \tau)|^2 = k^2 \tag{22.3}$$

defines the ambiguity function equation in one of the following forms:

$$\left.\begin{aligned} \Omega &= F_\Omega(\tau, k), \\ \tau &= F_\tau(\Omega, k). \end{aligned}\right\} \tag{22.4}$$

One of the most frequently used methods for selecting the value of k can be reduced to the following [14]:

(a) A cylinder of unit height, with generatrices perpendicular to the plane (Ω,τ), is constructed:

(b) the volume of the cylinder is equated to the volume of the "solid of ambiguity" in correspondence with condition (22.2);

(c) the value of k is found through equation

$$\iint_D d\Omega\, d\tau = 2\pi.$$

Then this value is substituted in (22.3) to find one of the equations (22.4) for the ambiguity diagram (in this case the domain of integration D depends on the condition $|\chi(\Omega,\tau)|^2 = k^2$).

Next, the ambiguity diagram is constructed in the form of curves in (Ω,τ) coordinates.

A simpler method consists of the following:

The ambiguity diagrams are found by sectioning the surface $|\chi(\Omega,\tau)|^2$ by planes parallel to (Ω,τ) at several levels $k_1, k_2, \ldots, k_n$, which are selected in terms of clarity of representation and convenience. Thus, we obtain something similar to curves representing equal heights (elevations) on topographic maps.

The simplest method for constructing ambiguity diagrams is to select one fixed value of k and compute the diagram using equation (22.3).

Figure 5 shows an example of an ambiguity function and diagram. In this case the ambiguity diagram was constructed by sectioning surface $|\chi(\Omega,\tau)|^2$ in a plane parallel to (Ω,τ) at level k^2.

Ambiguity diagrams can be used to evaluate the resolution of sonar systems with respect to time (that is, the

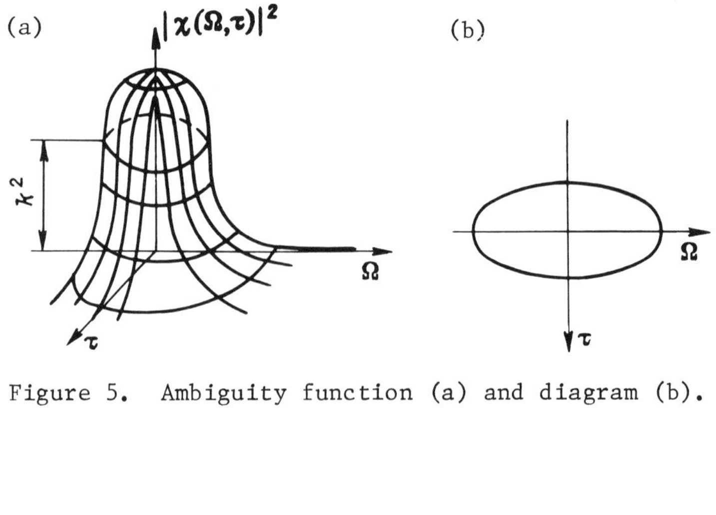

Figure 5. Ambiguity function (a) and diagram (b).

range to the object of detection) and frequency (that is, the rate of movement of the object). With suitable selection of the k values the resolution with respect to the coordinates indicated is defined by the points of intersection between the ambiguity diagram and the corresponding axes (Ω, τ). Thus, the values Ω_r and τ_r, which define the system's resolution with respect to a given form of the signal, are found by either of equations (22.4):

$$\Omega_r = F_\Omega(0, k), \qquad 0 = F_\Omega(\tau_r, k),$$

or

$$0 = F_\tau(\Omega_r, k), \qquad \tau_r = F_\tau(0, k).$$

As analysis has shown, by using different forms of emitted signals we can vary the values of Ω_r and τ_r over rather wide limits.

Parameters Ω_r and τ_r, which define the sonar system's resolution, can be interpreted as follows. On one hand these parameters characterize the effective number of alternatives (channels, independent readings) in range-velocity coordinates of the object of detection (see §54). On the other hand, τ_r characterizes the minimum differences in range and Ω_r characterizes the minimum difference in velocity that are "distinguishable" by the sonar system with respect to a given form of emitted signal. It is namely the ambiguity diagrams which represent these sonar system parameters.

In subsequent paragraphs of this chapter we will examine the characteristics of various types of emitted signals.

§23. Signals with Sinusoidal Occupation

In the case of using emitted signals with sinusoidal occupation, their complex envelope $Z_C(t)$ has the form

$$Z_C(t) = A(t),$$

where $A(t)$ is the real envelope characterizing the dependence of the signal's amplitude on time.

Let us find the ambiguity functions for several typical signals of this type.

In all cases let

$$\sup_t A^2(t) = 1, \qquad E_C = T_{\text{ef}},$$

where T_{ef} is the effective signal duration.

When this is taken into account, the ambiguity function $\chi(\Omega,\tau)$ takes the form

$$\chi(\Omega, \tau) = \frac{1}{T_{\text{ef}}} \int_{-\infty}^{\infty} A(t-\tau) A(t) \exp(j\Omega t)\, dt. \tag{23.1}$$

Let us examine a rectangular pulse for which

$$A(t) = \begin{cases} 1, & t \in [0, T], \\ 0, & t \bar{\in} [0, T], \end{cases} \tag{23.2}$$

where T is signal duration. We can easily see that

$$T_{\text{ef}} = T. \tag{23.3}$$

Substituting (23.2) and (23.1), we find that

$$\chi(\Omega, \tau) = \frac{1}{T} \int_{\tau}^{T} \exp(j\Omega t)\, dt = \frac{\exp(j\Omega T) - \exp(j\Omega\tau)}{j\Omega T} \tag{23.4}$$

$$= \frac{1}{\Omega T}\left[\exp\left(j\Omega T - j\frac{\pi}{2}\right) - \exp\left(j\Omega\tau - j\frac{\pi}{2}\right)\right], \qquad \tau \leqslant T.$$

Now let us find the complex conjugate value of $\chi(\Omega,\tau)$, viz.,

$$\chi^*(\Omega, \tau) = \frac{1}{T\Omega}\left[\exp\left(-j\Omega T + j\frac{\pi}{2}\right) - \exp\left(-j\Omega\tau + j\frac{\pi}{2}\right)\right]. \tag{23.5}$$

By definition,

$$|\chi(\Omega, \tau)|^2 = \chi(\Omega, \tau)\, \chi^*(\Omega, \tau). \tag{23.6}$$

Substituting (23.4) and (23.5) in (23.6) and carrying out the corresponding transformation, for the square of the modulus of the ambiguity function $|\chi(\Omega,\tau)|^2$ we find

$$|\chi(\Omega,\tau)|^2 = \left\{\frac{\sin\left[\frac{\Omega T}{2}\left(1 - \frac{|\tau|}{T}\right)\right]}{\Omega T/2}\right\}^2. \tag{23.7}$$

Let us examine a bell-shaped pulse, which is defined as a signal with envelope $A(t)$ described by a quadratic exponent:

$$A(t)=\exp\left[-\left(\frac{t}{t_0}\right)^2\right], \tag{23.8}$$

where t_0 is a parameter defining the signal's effective duration T_{ef}, in which case

$$T_{ef}=\left(\frac{\pi}{2}\right)^{1/2} t_0. \tag{23.9}$$

Using formulas (23.1), (23.8), and (23.9) to compute $|\chi(\Omega,\tau)|^2$ leads to the following expression:

$$|\chi(\Omega,\tau)|^2=\exp\left[-\frac{\pi}{2}\left(\frac{\tau}{T_{ef}}\right)^2-\frac{(\Omega T_{ef})^2}{2\pi}\right].$$

And, finally, the last example is that of an exponential pulse. For a signal of this type

$$A(t)=\begin{cases}\exp\left(-\frac{t}{t_0}\right), & t\in[0,\ \infty),\\ 0, & t\bar{\in}[0,\ \infty),\end{cases} \tag{23.10}$$

where in this case

$$T_{ef}=\frac{t_0}{2}. \tag{23.11}$$

After substituting (23.10) into (23.1) and considering (23.11), the square of the modulus of the ambiguity function $|\chi(\Omega,\tau)|^2$ turns out to be equal to

$$|\chi(\Omega,\tau)|^2=\frac{\exp\left(-\frac{|\tau|}{T_{ef}}\right)}{1+4(\Omega T_{ef})^2}. \tag{23.12}$$

Now we will find the ambiguity diagrams for some of the emitted signals examined above. We begin with (22.3) as the initial expression.

For a signal with a rectangular envelope, it follows from (23.7) that

$$\frac{\sin\left[\frac{\Omega T}{2}\left(1-\frac{|\tau|}{T}\right)\right]}{\Omega T/2}=k. \tag{23.13}$$

From this, solving equation (23.13) for $|\tau|/T$, we get

$$\frac{|\tau|}{T} = 1 - \frac{2}{\Omega T} \arcsin \frac{2k}{\Omega T}. \tag{23.14}$$

For a signal with an exponential envelope, according to (23.12)

$$\frac{\exp\left(-\frac{|\tau|}{T}\right)}{1 + 4(\Omega T_{\text{ef}})^2} = k^2. \tag{23.15}$$

Solving (23.15) for ΩT_{ef}, we have

$$\Omega T_{\text{ef}} = \frac{\exp\left(-\frac{|\tau|}{2T_{\text{ef}}} - \frac{k^2}{2}\right)}{2k}. \tag{23.16}$$

Analysis shows that a fixed relation between resolution with respect to time τ_r and resolution with respect to frequency Ω_r is typical for simple signals, in which case, for each type of envelope,

$$\tau_r \Omega_r = \text{const.} \tag{23.17}$$

This means that increasing signal duration always causes worsening of resolution with respect to τ and improvement with respect to Ω. It is obvious that simple signals of long duration are preferred for distinguishing moving objects of detection, while signals of short duration are preferred for distinguishing objects at close range.

Property (23.17) is a definite shortcoming of simple signals, which severely limits the area of their effective application in solving various statistical problems of sonar. More complicated signals do not have this shortcoming. We now turn to an examination of such signals.

§24. Frequency-Modulating Signals

In the case of using frequency-modulated signals, the phase $\Phi(t)$ must be found by the relationship

$$\Phi(t) = \int_0^t \omega(t')\, dt', \tag{24.1}$$

where $\omega(t')$ defines the law of frequency change -- that is the modulating function. In this case the ambiguity function $\chi(\Omega,\tau)$ is expressed by the relationship

$$\chi(\Omega, \tau) = \frac{1}{T_{ef}} \int_{-\infty}^{\infty} A(t-\tau) A(t) \exp[-j\Phi(t-\tau) + j\Phi(t)] \exp(j\Omega t)\, dt. \tag{24.2}$$

We will examine below a case of linear frequency modulation of a signal for two types of envelopes.

For a rectangular pulse with linear frequency modulation, envelope $A(t)$ is defined by relationship (23.2), while the frequency modulation law $\omega(t)$ has the form

$$\Omega(t) = \frac{\Delta\Omega_M}{T} t, \tag{24.3}$$

where $\Delta\Omega_M$ is the frequency deviation within the bounds of pulse duration T. According to (24.1) and (24.3)

$$\Phi(t) = \frac{\Delta\Omega_M}{2T} t^2, \tag{24.4}$$

and the square of the modulus of the ambiguity function, computed by formulas (23.2), (24.2), and (24.4), is expressed as

$$|\chi(\Omega, \tau)|^2 - \left\{ \frac{\sin\left[\frac{\Omega T + \Delta\Omega_M |\tau|}{2}\left(1 - \frac{|\tau|}{T}\right)\right]}{(\Omega T + \Delta\Omega_M |\tau|)/2} \right\}. \tag{24.5}$$

We note that parameter

$$\mu_M \equiv \frac{\Delta\Omega_M T}{2\pi}$$

defines the relative frequency deviation, which indicates the number of times the frequency deviation of the signal exceeds the width of the spectrum of its envelope. It is easy to see that by setting μ_M equal to zero we get a rectangular signal with sinusoidal occupation. In this case relationship (24.5) is the same as the expression (23.7) obtained earlier.

Next, let us examine a bell-shaped pulse with linear frequency modulation. In this case the envelope and phase of the signal are defined by relationships (23.8) and (24.4) respectively. Computing $|\chi(\Omega,\tau)|^2$ by formula (24.2) with a consideration for (23.5), (23.8), and (23.9), we get the following result:

$$|\chi(\Omega,\tau)|^2 = \exp\left[-\frac{\pi}{2}\left(\frac{\tau}{T_{ef}}\right)^2 - \frac{1}{2\pi}\left(\frac{\Delta\Omega_M|\tau|}{2\pi} + \Omega T_{ef}\right)^2\right]. \tag{24.6}$$

As an example, let us find the equation of the ambiguity diagram for a signal of this type. According to (22.3) and (24.6) we have

$$\exp\left[-\frac{\pi}{2}\left(\frac{\tau}{T_{ef}}\right)^2 - \frac{1}{2\pi}\left(\frac{\Delta\Omega_M|\tau|}{2\pi} + \Omega T_{ef}\right)^2\right] = k^2,$$

whence

$$\frac{\pi}{2}\left(\frac{\tau}{T_{ef}}\right)^2 + \frac{1}{2\pi}\left(\frac{\Delta\Omega_M|\tau|}{2\pi} + \Omega T_{ef}\right)^2 = 2\ln\frac{1}{k}. \tag{24.7}$$

The ambiguity diagram represents the equation of an ellipse, the axis of which is rotated a certain angle relative to the coordinate axes, depending on the value of parameter

$$\mu_M = \frac{\Delta\Omega_M T_{ef}}{2\pi}$$

As a comparison of the forms of ambiguity diagrams encountered with frequency-modulated signals, Figure 6 shows three diagrams constructed at various values of parameter μ_M.

An analysis of the expression obtained indicates that frequency-modulated emitted signals permit us to attain heightened resolution with respect to time τ, in which case this effect increases as the frequency deviation grows. At the same time the resolution with respect to coordinate Ω -- that is, with respect to frequency, remains practically unchanged and depends on the effective signal duration, which does not depend on the nature of frequency deviation. This interesting fact can be utilized in solving various statistical problems in sonar.

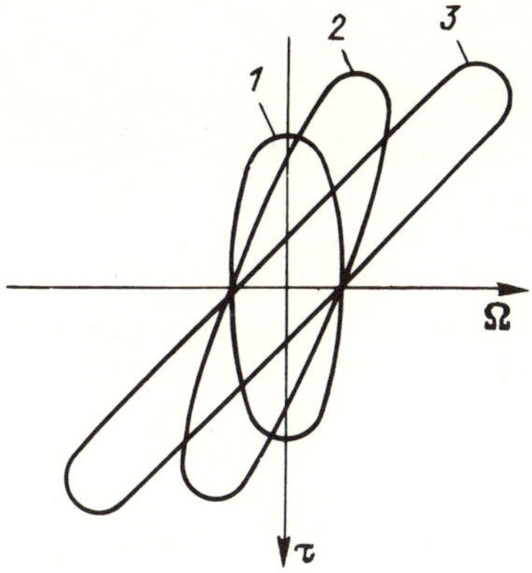

Figure 6. Ambiguity diagrams for a frequency-modulated signal. Curve 1, at $\mu_M = 0$; curve 2, at $\mu_{M1} > 0$; curve 3, at $\mu_{M2} > \mu_{M1}$.

§25. Noiselike Signals

The complex envelope $Z_C(t)$ for noiselike signals is a random function

$$Z_C(t) = E(t) \exp [j \Psi(t)],$$

where $E(t)$ and $\Psi(t)$ are, in the general case, the fluctuating envelope and the phase of the signal respectively. Therefore the ambiguity function

$$\chi(\Omega, \tau) = \frac{1}{\langle E_C \rangle} \int_{-\infty}^{\infty} E(t-\tau) E(t) \exp[-j\Psi(t-\tau) + j\Psi(t)] \exp(j\Omega t)\, dt \tag{25.1}$$

is also a random function, the value of which would change from one representation of the signal to another; (the variable $\langle E_C \rangle$ in formula (25.1) is the average value of the signal's energy). In examining noiselike signals we can refer to some mean square modulus of the ambiguity function $\langle |\chi(\Omega,\tau)|^2 \rangle$.

Considering that

$$\langle |\chi(\Omega, \tau)|^2 \rangle = \langle \chi(\Omega, \tau) \chi^*(\Omega, \tau) \rangle,$$

and also taking (25.1) into account, we get

$$\langle|\chi(\Omega,\tau)|^2\rangle=\frac{1}{\langle E_C\rangle^2}\int_{-\infty}^{\infty}\int_{-\infty}^{\infty}\langle E(t'-\tau)\,E(t')\,E(t''-\tau)\,E(t'')\times$$

$$\times\exp[-j\Psi(t'-\tau)+j\Psi(t')+j\Psi(t''-\tau)-j\Psi(t'')]\rangle\exp[-j\Omega(t'-t'')]\,dt'\,dt''. \tag{25.2}$$

The noiselike signal class can contain three subclasses, namely random amplitude modulation, random frequency (or phase) modulation, and formation of a signal in the form of an interval of noise of some particular duration. In the first case, when only the envelope $E(t)$ is random and when $\Psi(t) = \Phi(t)$ is a deterministic function, only those factors of integrand (25.2) that contain the envelope are subjected to statistical averaging. In this case we get

$$\langle|\chi(\Omega,\tau)|^2\rangle=\frac{1}{\langle E_C\rangle^2}\int_{-\infty}^{\infty}\int_{-\infty}^{\infty}\langle E(t'-\tau)\,E(t')\,E(t''-\tau)\,E(t')\rangle\times$$

$$\times\exp[-j\Phi(t'-\tau)+j\Phi(t')+j\Phi(t''-\tau)-j\Phi(t'')]\exp[j\Omega(t'-t'')]\,dt'\,dt''.$$

In the second case involving random frequency (or phase) modulation, when $E(t) = A(t)$ is a deterministic function, we have

$$\langle|\chi(\Omega,\tau)|^2\rangle=\frac{1}{\langle E_C\rangle^2}\int_{-\infty}^{\infty}\int_{-\infty}^{\infty}A(t'-\tau)\,A(t')\,A(t''-\tau)\,A(t'')\times$$

$$\times\langle\exp[-j\Psi(t'-\tau)+j\Psi(t')+j\Psi(t''-\tau)-j\Psi(t'')]\rangle\exp[j\Omega(t'-t'')]\,dt'\,dt''.$$

And, finally, in the last case the signal's complex envelope $Z_C(t)$ is represented in the form

$$Z_C(t)=\begin{cases}X(t), & t\in[0,\ T],\\ 0, & t\,\bar{\in}\,[0,\ T]\end{cases}$$

and is an interval of noise $X(t)$ with duration T. In this case

$$\langle|\chi(\Omega,\tau)|^2\rangle=\frac{1}{\langle E_C\rangle^2}\int_{0}^{T-\tau}\int_{0}^{T-\tau}\langle X(t'-\tau)\,X^*(t')\,X^*(t''-\tau)\,X(t'')\rangle\times$$

$$\times\exp[j\Omega(t'-t'')]\,dt'\,dt'',\qquad|\tau|<T. \tag{25.3}$$

At this point we will examine only the last emitted signal subclass and assume that $X(t)$ is a stationary Gaussian random process.

Let us determine the mean square of the ambiguity function's modulus $\langle|\chi(\Omega,\tau)|^2\rangle$ as expressed in (25.3) above. As we know, in view of the fact that $X(t)$ is a stationary Gaussian random process, for the four-dimensional moment $\langle X(t' - \tau)X^*(t')X^*(t'' - \tau)X(t'')\rangle$ in (25.3) we have

$$\langle X(t'-\tau)X^*(t')X^*(t''-\tau)X(t'')\rangle = B_{XX^*}(\tau)B_{X^*X}(\tau) + \\ + B_{XX^*}(t'-t'')B_{X^*X}(t'-t'') + B_{XX}(t'-t''-\tau)B_{X^*X^*}(t'-t''+\tau); \tag{25.4}$$

$$\left.\begin{aligned} &B_{XX^*}(\tau)B_{X^*X}(\tau) = 4B_x^2(\tau);\\ &B_{XX^*}(t'-t'')B_{X^*X}(t'-t'') = 4B_x^2(t'-t'');\\ &B_{XX}(t'-t''-\tau)B_{X^*X^*}(t'-t''+\tau) =\\ &\qquad = 4B_x(t'-t''-\tau)B_x(t'-t''+\tau), \end{aligned}\right\} \tag{25.5}$$

where

$$B_x(\tau) \equiv \langle x(t+\tau)x(t)\rangle$$

is an autocorrelation function of the real process $x(t)$, obtained from the expression

$$X(t) = x(t) - j\tilde{x}(t).$$

We also note that in this case the average energy of the emitted signal is defined as

$$\langle E_C\rangle = \int_0^T \langle|X(t)|^2\rangle\,dt = I_X T = 2I_x T, \tag{25.6}$$

where $I_X = \langle|X(t)|^2\rangle$ is the average intensity of the complex envelope of the noise occupation and $I_x = \langle x^2(t)\rangle$ is the average intensity of the real process $x(t)$.

Now we can write the value of the mean square of the ambiguity function's modulus $\langle|\chi(\Omega,\tau)|^2\rangle$. On the basis of relationships (25.3)-(25.6) we get

$$\langle|\chi(\Omega,\tau)|^2\rangle = \frac{1}{I_x^2 T^2}\left\{B_x^2(\tau)\int_0^{T-\tau}\int_0^{T-\tau}\exp\left[j\Omega\left(t'-t''\right)\right]dt'\,dt'' + \right.$$

$$+\int_0^{T-\tau}\int_0^{T-\tau} B_x^2\left(t'-t''\right)\exp\left[j\Omega\left(t'-t''\right)\right]dt'\,dt'' + \int_0^{T-\tau}\int_0^{T-\tau} B_x\left(t'-t''-\tau\right)\times$$

$$\left.\times B_x\left(t'-t''+\tau\right)\exp\left[j\Omega\left(t'-t''\right)\right]dt'\,dt''\right\},\quad |\tau|\leqslant T. \qquad (25.7)$$

The general expression (25.7) for $\langle|\chi(\Omega,\tau)|^2\rangle$ can be simplified. In the first place, we consider that

$$\int_0^{T-\tau}\int_0^{T-\tau}\exp\left[j\Omega\left(t'-t''\right]dt'\,dt'' = \frac{4}{\Omega^2}\sin^2\left[\frac{\Omega T}{2}\left(1-\frac{\tau}{T}\right)\right].$$

Secondly, we reduce the double integral (25.7) to a single integral. Then expression (25.7) for $\langle|\chi(\Omega,\tau)|^2\rangle$ takes the form:

$$\langle|\chi(\Omega,\tau)|^2\rangle = \frac{1}{I_x^2}\left\{B_x^2(\tau)\,\frac{\sin\left[\frac{\Omega T}{2}\left(1-\frac{\tau}{T}\right)\right]}{(\Omega T/2)^2} + \right.$$

$$+2\,\frac{1}{T}\left(1-\frac{\tau}{T}\right)\int_0^{T-\tau}\left(1-\frac{t}{T-\tau}\right)B_x^2(t)\exp\left(j\Omega t\right)dt +$$

$$\left.+\int_0^{T-\tau}\int_0^{T-\tau}B_x\left(t'-t''-\tau\right)B_x\left(t'-t''+\tau\right)\exp\left[j\Omega\left(t'-t''\right)\right]dt'\,dt''\right\}. \qquad (25.8)$$

For subsequent computations we must assign a particular form to the autocorrelation function $B_x(t)$ for the noise occupation of the emitted signal.

Here, then, is the general result of using noiselike signals: When the complexity factor of such signals is increased, resolution with respect to time τ increases and is defined by the correlation interval of noise occupation. At the same time resolution with respect to frequency Ω depends weakly on noise occupation and is defined by the effective signal duration. This result is natural from the standpoint of the frequency correlation of emitted noise: The interval of frequency correlation (frequency resolution) for nonstationary random processes is inversely proportional to the nonstationary interval of the process, and it is precisely this interval of the emitted signal that defines its duration.

CHAPTER V

ECHO SIGNALS

§26. Classification of Objects of Research and the Conditions under Which They Are Observed

Any sort of classification, including the one being examined in this paragraph, requires a definition of the classification indices of importance to the problem at hand. Since we are interested in developing probability models of echo signals, we should use the distortions to which emitted signals are subjected as they propagate in a water medium and when they reflect from objects of research as the main classification index. Thus, our examination reduces to studying the differences between forms of the echo signal and forms of the emitted signal.

We will use constructive models of echo signals [20, 22,34,51,67], in which the signal will be represented as an additive-multiplicative mixture of a signal of known form

Table 1

Underwater observation conditions	Nature of distortions in echo signals
Homogeneous unbounded medium	None
Reflection of signal from smooth boundaries	Presence of several elementary echo signals
Scattering of signal on statistically uneven boundaries	Presence of additive random component in echo signal
Multiple-beam propagation of signal in the presence of medium irregularities and unevennesses	Presence of several elementary echo signals and an additive random component
Temporal variability of medium characteristics	Random amplitude and frequency modulation

Table 2

Objects of detection	Nature of distortions in echo signal for:		
	Nonmoving objects	Objects moving straight and uniformly	Objects moving with random relative velocity
"Ideal" spherical reflector	- - -	Constant change in echo signal's time scale*	Random frequency modulation
Object with several "highlights"	Presence of several elementary echo signals	Presence of several elementary echo signals with constant changes in time scale	Presence of several elementary echo signals, random amplitude and frequency modulation
Object with one highlight and a scattering surface	Presence of additive random component	Presence of additive random component in echo signal and constant change in echo signal time scale	Presence of random additive component in echo signal and random frequency modulation
Groups of objects statistically distributed in space	Presence of many elementary echo signals with random amplitudes and initial phases	Presence of many elementary echo signals with random amplitudes, initial phases, and changes in time scale	Presence of many elementary echo signals with random amplitudes, initial phases, and random amplitude modulation

*Change in time scale implies "compression" or "stretching" of echo signals in time in response to Doppler effect.

and random functions defining distortion factors (such models had been examined in §17). In this case distortions in the echo signal (relative to the undistorted emitted signal) will be defined as additive, multiplicative, and combined components.

Tables 1 and 2 provide classifications of the underwater observation conditions and the objects of research from the standpoints defined above. Table 1 is set up on the assumption that the object of detection does not contribute any sort of distortions to the echo signal, while Table 2 is set up on the assumption that the water medium is homogeneous and boundless -- that is, that it does not contribute distortions to the echo signal as well. Such an approach follows naturally from the point of view of the sonar model discussed in §2: We isolate the effects of operator M, the water medium, and operator T, which describes the properties of the object of research (detection). Such isolation subsequently enables us to examine generalized constructive models of echo signals in which both factors (the water medium and the object) are no longer differentiated in regard to their effects.

The distortions noted in the tables above enable us to develop a whole series of constructive probability models of echo signals. The following are the simplest of such models:

echo signal of known form with constant lag and constant change in time scale (effect of movement of the object of detection);

echo signal with fluctuating amplitude;

echo signal with random additive component;

echo signal with random amplitude modulation;

echo signal with random phase modulation;

echo signal with random frequency modulation;

echo signal in the form of a sum of elementary echo signals with fixed or random parameters.

In discussing the echo signal models we will not distinguish between distortions arising in response to the joint effect of underwater observation conditions and the objects of research. Of course, such an assumption would make our examination of echo signal properties incomplete, since distortions of the combined type may be observed in real echo signals. Nevertheless, an analysis of these models of echo signals makes great methodological sense because we can make a clear distinction among the effects of fully defined distortions on echo signal properties.

§27. Representation of Echo Signals* and Their Probability Characteristics

Analogously with our analysis of the characteristics of emitted signals $C(t)$, where we dealt with their complex envelope $Z_C(t)$, in this case we will refer to the complex envelope $Z_S(t)$ of echo signals $S(t)$. Obviously, even if the emitted signal is determinate the echo signal is described by a random function. This is connected with the random distortions the form of the echo signal experiences and with the uncertainty about its parameters stemming from *a priori* ignorance of the properties of the objects of research and the adoption of particular hypotheses regarding their properties (range to the object, its rate of movement, its reflecting capability).

Selection of the types of probability characteristics $\Theta(l \mid m)$ corresponding to the models m of the processes (echo signals) under consideration is important in describing echo signals $Z_S(t)$ in terms of probability.

An echo signal is described fully by its n-dimensional probability density

$$W_n(Z_C; l \mid m) = W(Z_{C1}, Z_{C2}, \ldots, Z_{Cn}; l \mid m). \quad (27.1)$$

As a rule, however, we find that *a priori* information and substantiated hypotheses regarding the properties of the objects of research are not enough to obtain the multidimensional characteristic (27.1). In addition, it is practically impossible to test experimentally the adequacy

---

*See footnote, p. 109 (D.M.)

of these models with respect to the objects of research in terms of n-dimensional probability densities in view of the lack of experimental data and the difficulties in making statistical measurements and in interpreting the results. For this reason we find it convenient to study the numerical characteristics of multidimensional distributions, particularly their moment functions (this is often the only approach possible). In some cases one-dimensional probability densities are studied directly, but most often they are obtained by approximating them by convergent series (Edgeworth, Gram-Charlier [12], and others) with respect to moment functions or semi-invariants.

We note that in a number of cases the probability distributions of sonar signals are non-Gaussian [2,14,16, 21,22] -- that is, they cannot be described by Gaussian laws of probability distribution. In such cases the study of moment functions is reduced to defining these functions up to rather high orders of magnitude. It should be kept in mind, however, that in accordance with the echo signal models being developed the signals can be thought of either as originating from Gaussian signals or (in some cases) as being adequately close to Gaussian signals. In this case correlational or spectral descriptions may be an adequate level of probability description. We will subsequently examine the characteristics of correlational echo signals.

If we are to satisfy the requirements for analyzing and synthesizing sonar systems, and if we are to obtain initial data to be used in solving inverse problems, then we must study the current-t probability characteristics of the echo signals (echo signals are nonstationary random processes in view of their pulsed nature).

By definition, the current-t autocorrelation function $B(t,\tau)$ of echo signal $Z_S(t)$ has the form

$$B(t, \tau) \equiv \langle Z_S(t-\tau) Z_S^*(t) \rangle. \qquad (27.2)$$

Correspondingly, the current-t energy spectrum $G(t,\omega)$ can be defined as follows:

$$G(t, \omega) \equiv \int_{-\infty}^{\infty} B(t, \tau) \exp(-j\omega\tau)\, d\tau. \qquad (27.3)$$

A normalized statistical expression for the short-term frequency-time cross-correlation function (or cross-ambiguity function) of the complex envelopes of the emitted signal and echo signal is an important characteristic of the echo signal. This expression, $\hat{\varkappa}(\Omega,\tau)$, is defined as follows:

$$\hat{\varkappa}(\Omega, \tau) \equiv \frac{1}{(E_C \langle E_S \rangle)^{1/2}} \int_{-\infty}^{\infty} Z_C(t-\tau) Z_S^*(t) \exp(j\Omega t)\, dt, \qquad (27.4)$$

where

$$E_C = \int_{-\infty}^{\infty} |Z_C(t)|^2\, dt \qquad (27.5)$$

is the energy of the emitted signal, and

$$\langle E_S \rangle = \int_{-\infty}^{\infty} |Z_S(t)|^2\, dt \qquad (27.6)$$

is the average energy of the echo signal.

In a certain sense, the quantity $\hat{\varkappa}(\Omega,\tau)$ can be thought of as a normalized characteristic of similarity (or closeness) for signals $Z_C(t)$ and $Z_S(t)$. In fact, according to definition (27.4), the greater the match between the form of the emitted signal and that of the echo signal, the greater the value of $\hat{\varkappa}(\Omega,\tau)$ should be. In particular, if these signals match in form precisely and $Z_C(t) = Z_S(t)$, then $\hat{\varkappa}(\Omega,\tau)$ is equal to the ambiguity function (see §22). In general appearance, the characteristic (27.4) recalls the appearance of the ambiguity function. In contrast to it, however, $\hat{\varkappa}(\Omega,\tau)$ is a statistical quantity -- that is, it is a random function -- because the form of the echo signal $Z_S(t)$, the parameters, are always partially unknown. Therefore, we are interested in the probability characteristics of the quantity $\hat{\varkappa}(\Omega,\tau)$, chiefly the following:

(a) Mathematical expectation

$$m_{\varkappa}(\Omega, \tau) \equiv \frac{1}{(E_C \langle E_S \rangle)^{1/2}} \int_{-\infty}^{\infty} Z_C(t-\tau) \langle Z_S^*(t) \rangle \exp(j\Omega t)\, dt; \qquad (27.7)$$

(b) Variance of fluctuations

$$d_\varkappa(\Omega, \tau) \equiv \langle |\hat{\varkappa}(\Omega, \tau)|^2\rangle - |m_\varkappa(\Omega, \tau)|^2 =$$
$$= \frac{1}{E_C \langle E_S\rangle} \int_{-\infty}^{\infty}\int_{-\infty}^{\infty} Z_C(t' - \tau) Z_C^*(t'' - \tau) \langle Z_S^*(t') Z_S(t'')\rangle \times$$
$$\times \exp[j\Omega(t' - t'')]\, dt'\, dt'' - |m_\varkappa(\Omega, \tau)|^2; \qquad (27.8)$$

(c) Correlation function for the cross-ambiguity functions

$$B_\varkappa(\Omega_1, \Omega_2; \tau_1, \tau_2) \equiv \langle \hat{\varkappa}(\Omega_1, \tau_1)\hat{\varkappa}^*(\Omega_2, \tau_2)\rangle -$$
$$- m_\varkappa(\Omega_1, \tau_1) m^*(\Omega_2, \tau_2) = \frac{1}{E_C\langle E_S\rangle}\int_{-\infty}^{\infty}\int_{-\infty}^{\infty} Z_C(t' - \tau_1) \times$$
$$\times Z_C^*(t'' - \tau_2)\langle Z_S^*(t') Z_S(t'')\rangle \exp(j\Omega_1 t' -$$
$$- j\Omega_2 t'')\, dt'\, dt'' - m_\varkappa(\Omega_1, \tau_1) m^*(\Omega_2, \tau_2). \qquad (27.9)$$

In principle we can pose the problem of studying the probability distribution of the quantity $\hat{\varkappa}(\Omega,\tau)$ in the frequency-time domain. These probability distributions can be found for different mathematical models of echo signals by approximation methods, after computing the corresponding moment functions. However, although this can be done, the volume of computations necessary is rather large. If we consider that the distribution of the quantity $\hat{\varkappa}(\Omega,\tau)$ is Gaussian or has Gaussian origins for a whole series of cases of practical interest, then characteristics (27.7)-(27.9) are more than adequate to describe the probability properties of this quantity.

In subsequent paragraphs we will examine different constructive models of echo signals and define some of their probability characteristics.

§28. The Model of an Echo Signal of Known Form with Constant Lag and Change in Time Scale

Let us examine the simplest model of an echo signal, in which the signal's form is known and in which it differs from an emitted signal with complex envelope $Z_C(t)$ by a lag in time and by a change in the time scale owing to movement of the object of research.

We first consider the lag of the echo signal. Starting with the general representation

$$Z_C(t) = A(t) \exp[-j\Phi(t)]$$

we write the complex envelope $Z_S(t)$ of the echo signal in the following form:

$$Z_S(t) = A_S A(t - \tau_S) \exp[-j\Phi(t - \tau_S)],$$

where A_S is the amplitude of the echo signal, τ_S is the time lag of the echo signal relative to the emitted signal, and functions $A(t)$ and $\Phi(t)$ define the amplitude envelope and the instantaneous phase of the emitted signal and the echo signal.

Now we consider the effect of movement of the object of detection [65].

We introduce parameter

$$\nu = \frac{2v}{c} \cos\theta,$$

which describes the relative velocity of the object of research (v is its absolute velocity, c is the rate of propagation of the acoustic signal in the water medium, and Θ is the angle between the direction of movement of the object and a line between the object and the sonar system). There is a dual effect of the movement of the object of detection on the form of the echo signal, namely,

(1) its spectrum is transformed in frequency by an amount $\nu\omega_0$, where ω_0 is the central frequency of the emitted signal's spectrum;

(2) the time scale of function $Z_C(t)$ changes by a factor $1 + \nu$.

Considering the above, for $Z_S(t)$ we get

$$Z_S(t) = A_S A[t(1 + \nu) - \tau_S) \times$$
$$\times \exp\{-j\Phi[t(1 + \nu) - \tau_S] \exp[-j\nu\omega_0(t - \tau_S)]\}. \quad (28.1)$$

It is interesting to note that the effect of the movement of objects of detection on the echo signal form differs in magnitude in radar and sonar. While in radar, the parameter ν has an approximate value $\nu_r = 10^{-6}$, in sonar $\nu_s = 10^{-2}$. This means that in radar a change in time scale by $1 + \nu$ times can be ignored, while in sonar this cannot be done in many cases.

We write the echo signal expression (28.1) in the short form:

$$Z_S(t) = A_S Z_C[t(1+\nu) - \tau_S] \exp[-j\Omega_S(t - \tau_S)]. \tag{28.2}$$

We define the quantity $\hat{\varkappa}(\Omega,\tau)$. According to (27.4) and (28.2) now,

$$\hat{\varkappa}(\Omega, \tau) = \frac{A}{(E_C \langle E_S \rangle)^{1/2}} \int_{-\infty}^{\infty} Z_C(t - \tau) Z_C^*[t(1+\nu) - \tau_S] \exp[j\nu\omega_0(t - \tau_S) + j\Omega t]\, dt. \tag{28.3}$$

When $\nu = 0$ -- that is, when the object of detection has zero relative velocity, it follows from (28.3) that

$$\hat{\varkappa}(\Omega, \tau) = \chi(\Omega, \tau - \tau_S) \exp(j\Omega\tau_S).$$

In other words, characteristic $\hat{\varkappa}(\Omega,\tau)$ matches correctly within a factor $\exp(j\Omega\tau_S)$ in its form to the emitted signal's ambiguity function $\chi(\Omega,\tau)$ shifted in time by τ_S relative to the moment of emission $\tau_S = 0$.

If the emitted signals are simple -- that is, if

$$Z_C(t) = A(t), \tag{28.4}$$

then on substituting (28.4) into (28.3) we get

$$\hat{\varkappa}(\Omega, \tau) = \frac{A_S}{(E_C \langle E_S \rangle)^{1/2}} \int_{-\infty}^{\infty} A(t - \tau) A[t(1+\nu) - \tau_S] \exp[j\nu\omega_0(t - \tau_S) + j\Omega t]\, dt. \tag{28.5}$$

An analysis of expression (28.5) shows that at $\nu \neq 0$, the characteristic $\hat{\varkappa}(\Omega,\tau)$ does not match in form with the ambiguity function $\chi(\Omega,\tau)$. However, the difference is found to be insignificant and involves only a change in the echo signal's effective duration by a factor $1 + \nu$.

Movement of the objects of detection has a more significant influence when complicated signals are employed, since the Doppler effect causes not only a shift in the average frequency of the echo signal's spectrum and a change in the signal's effective duration, but transforms the law of its phase (or frequency) modulation. In this case, these transformations become more significant as the ratio between the effective width of the signal's spectrum and its average frequency grows.

As an example, let us examine the emission of a signal with linear frequency modulation, of which the phase $\Phi(t)$ is defined by the ratio

$$\Phi(t) = \frac{\Delta\omega_{ef}}{2T_{ef}} t^2,$$

where $\Delta\omega_{ef}$ and T_{ef} are the effective frequency deviation and duration of the signal, respectively. It is easy to see that the phase $\Phi_S(t)$ of the echo signal is defined in this case, with a consideration for movement of the object of detection, by the expression

$$\Phi_S(t) = \frac{\Delta\omega_{ef\nu}}{2T_{ef}} t^2,$$

where

$$\Delta\omega_{ef\nu} = (1+\nu)^2 \Delta\omega_{ef}$$

is a certain effective frequency deviation in the echo signal. In general, inasmuch as

$$\Phi(t) = \int_0^t \Omega(t')\, dt',$$

where the function $\Omega(t')$ defines the law of frequency modulation, for phase $\Phi_S(t)$ of the echo signal we have

$$\Phi_S(t) = \int_0^{t(1+\nu)} \Omega_S(t')\, dt',$$

where $\Omega_S(t')$ is the law of frequency modulation in the echo signal.

It should be kept in mind that the distortions in the echo signal examined above, caused by movement of the object of detection, can be defined for each expected value of ν -- that is, they are actually known *a priori* and are controllable. Therefore, generally speaking, we can compute the characteristic of signal similarity $\hat{\varkappa}(\Omega,\tau)$, for each ν after correspondingly replacing the complex envelope

$Z_C(t)$ by a certain function $Z_{C\nu}(t)$ which takes account of the expected distortions of the type examined -- that is,

$$Z_{C\nu}(t) - Z_C[t(1+\nu)].$$

Obviously, in this case characteristic $\hat{\varkappa}(\Omega,\tau)$ can be defined not by formula (28.3) but rather by

$$\hat{\varkappa}(\Omega, \tau) \equiv \frac{A_S}{(E_C\langle E_S\rangle)^{1/2}} \int_{-\infty}^{\infty} Z_C[t(1+\nu)-\tau] \times$$
$$\times Z_C^*[t(1+\nu)-\tau_S] \exp[-j\nu\omega_0(t-\tau_S)+j\Omega t]\, dt. \quad (28.6)$$

It follows from (28.6), in particular, that

$$\hat{\varkappa}(\Omega, \tau) = \chi(\Omega-\nu\omega_0, \tau-\tau_S)\exp\left(j\frac{\nu^2}{1+\nu}\omega_0\tau_S + j\frac{\Omega\tau_S}{1+\nu}\right). \quad (28.7)$$

Relationship (28.7) shows that if signal $Z_C(t)$ is replaced by a corrected signal $Z_{C\nu}(t)$, then the characteristic $\hat{\varkappa}(\Omega,\tau)$ matches the emitted signal's ambiguity function correctly to within the factor

$$\exp\left(j\frac{\nu^2}{1+\nu}\omega_0\tau_S + j\frac{\Omega\tau_S}{1+\nu}\right).$$

It is obvious that $|\hat{\varkappa}(\Omega,\tau)|^2 = |\chi(\Omega,\tau)|^2$. For this reason we will subsequently not take account of the effect of these distortions on the form of the echo signal -- that is, we will assign $\nu = 0$ and $\tau_S = 0$, assuming that the characteristic $\hat{\varkappa}(\Omega,\tau)$ is computed with a consideration for selection of the corresponding complex envelope of the emitted signal, in which the distortions examined above are necessarily taken into account.

§29. Echo Signal with Fluctuating Amplitude

For the case of an echo signal with fluctuating amplitude,

$$Z_S(t) = A_S Z_C(t),$$

where A_S is the echo signal's random amplitude. Let us construct the expression for the function $\hat{\varkappa}(\Omega,\tau)$:

$$\hat{\varkappa}(\Omega, \tau) = \frac{A_S}{(E_C \langle E_S \rangle)^{1/2}} \int_{-\infty}^{\infty} Z_C(t-\tau) Z_C^*(t) \exp(j\Omega t)\, dt. \tag{29.1}$$

It is easy to see that the echo signal's average energy $\langle E_S \rangle$ equals

$$\langle E_S \rangle = \langle A_S^2 \rangle E_C. \tag{29.2}$$

Substituting (29.2) in (29.1) we find that

$$\hat{\varkappa}(\Omega, \tau) = \frac{A_S}{(\langle A_S^2 \rangle)^{1/2}} \chi(\Omega, \tau), \tag{29.3}$$

where $\chi(\Omega,\tau)$ is the ambiguity function.

In accordance with (27.7) and (29.3) the average value of $\chi(\Omega,\tau)$ is found to be:

$$m_\varkappa(\Omega, \tau) = \frac{\langle A_S \rangle}{(\langle A_S^2 \rangle)^{1/2}} \chi(\Omega, \tau). \tag{29.4}$$

The variance $d_\varkappa(\Omega,\tau)$ is found to be equal, according to (27.8), (29.3), and (29.4), to

$$d_\varkappa(\Omega, \tau) = \left(1 - \frac{\langle A_S \rangle^2}{\langle A_S^2 \rangle}\right) |\chi(\Omega, \tau)|^2. \tag{29.5}$$

Thus, fluctuations in the intensity of the echo signal do not change the form of the average value, $m_\varkappa(\Omega,\tau)$, which matches the ambiguity function $\chi(\Omega,\tau)$. However, they do cause the variance $d_\varkappa(\Omega,\tau)$ to be other than zero, its value depending on the type of distribution of the random amplitudes A_S.

Let us evaluate the coefficient of variation for characteristic $\hat{\varkappa}(\Omega,\tau)$, defined by

$$\gamma_\varkappa(\Omega, \tau) \equiv \frac{d_\varkappa^{1/2}(\Omega, \tau)}{|m_\varkappa(\Omega, \tau)|} \tag{29.6}$$

and introduce a corresponding coefficient of variation for echo signal amplitude A_S:

$$\gamma_A \equiv \frac{d_A^{1/2}}{m_A}, \tag{29.7}$$

where

$$m_A = \langle A_S \rangle, \qquad d_A = \langle A_S^2 \rangle - \langle A_S \rangle^2.$$

Using relationships (29.4)-(29.7) we get here

$$\gamma_\varkappa(\Omega, \tau) = \gamma_A. \tag{29.8}$$

§30. Echo Signal Consisting of an Additive Mixture of a Signal of Known Form and a Random Component

In the instance of an echo signal consisting of an additive mixture of a signal of known form and a random component, the echo signal's complex envelope $Z_S(t)$ is defined by the relationship

$$Z_S(t) = Z_C(t) + X(t), \tag{30.1}$$

in which case the first term coincides in form with the emitted signal and $X(t)$ is a random stationary process with an average value equal to zero: $\langle X(t) \rangle = 0$. The random component $X(t)$ may appear in the echo signal as a result of, for example, scattering of acoustic waves off of a rough surface on the object of research, or scattering of these waves off of irregularities in the water medium or off of its boundaries. It should be noted that the echo signal's additive component $X(t)$ is not marine reverberation, which also stems from scattering of acoustic waves off of irregularities in the water medium or off of its boundaries. Here is the difference: while marine reverberation arises at the point of reception as a result of scattering of the primary signal off of irregularities in the water medium and off of its boundaries, this additive component arises either as a result of scattering of the primary signal off of rough surfaces on the object of research or as a result of scattering of the echo signal as it propagates. Thus, the additive component $X(t)$ exists only when an object of research is present, while marine reverberation always exists irrespective of the presence or absence of an object of research. In addition to this, we should keep in mind that in terms of its probability properties, $X(t)$ may be a "reverberation-like" additive component.

For the quantity $\hat{\varkappa}(\Omega, \tau)$ of interest to us, we have here

$$\hat{\varkappa}(\Omega, \tau) = \frac{1}{(E_C \langle E_S \rangle)^{1/2}} \int_{-\infty}^{\infty} Z_C(t-\tau)\left[Z_C^*(t) + X^*(t)\right] \exp(j\Omega t)\, dt,$$

that is,

$$\hat{\varkappa}(\Omega, \tau) = \left(\frac{E_C}{\langle E_S \rangle}\right)^2 \chi(\Omega, \tau) + \frac{1}{(E_C \langle E_S \rangle)^{1/2}} \int_{-\infty}^{\infty} Z_C(t-\tau) X^*(t) \exp(j\Omega t)\, dt. \tag{30.2}$$

We define the average energy $\langle E_S \rangle$ of the echo signal as

$$\langle E_S \rangle = \int_{-\infty}^{\infty} \langle | Z_S(t) |^2 \rangle\, dt. \tag{30.3}$$

Taking (30.1) into account, we see that (30.3) leads to

$$\langle E_S \rangle = \int_{-\infty}^{\infty} \langle \left[Z_C(t) + X(t)\right]\left[Z_C^*(t) + X^*(t)\right] \rangle\, dt.$$

But in view of the statistical independence of processes $Z_C(t)$ and $X(t)$, we have

$$\langle Z_C(t) X(t) \rangle = \langle Z_C^*(t) X(t) \rangle = \langle Z_C(t) X^*(t) \rangle = 0. \tag{30.4}$$

Designating by

$$d_X = \langle X(t) X^*(t) \rangle$$

the (ω) variance of process $X(t)$, and considering that the effective duration of the emitted signals $Z_C(t)$ being examined is T_{ef}, for $\langle E_S \rangle$ we get

$$\langle E_S \rangle = E_C + d_X T_{ef}. \tag{30.5}$$

We introduce parameter

$$Q = \frac{E_C}{d_X T_{ef}}, \tag{30.6}$$

which characterizes the ratio between the energy of the deterministic and random parts of the echo signal. Then, taking (30.5) and (30.6) into account, we derive the following from relationship (30.2):

$$\hat{\varkappa}(\Omega, \tau) = \left(\frac{Q}{1+Q}\right)^{1/2} \chi(\Omega, \tau) + \frac{1}{E_C}\left(\frac{Q}{1+Q}\right)^{1/2} \int_{-\infty}^{\infty} Z_C(t-\tau) X^*(t) \exp(j\Omega t)\, dt. \tag{30.7}$$

As in the previous case, we are interested in the average value of the function $\hat{\varkappa}(\Omega,\tau)$

$$m_\varkappa(\Omega, \tau) = \langle \hat{\varkappa}(\Omega, \tau)\rangle$$

and its variance

$$d_\varkappa(\Omega, \tau) = I_\varkappa(\Omega, \tau) - |m_\varkappa(\Omega, \tau)|^2, \tag{30.8}$$

where

$$I_\varkappa(\Omega, \tau) = \langle |\hat{\varkappa}(\Omega, \tau)|^2\rangle. \tag{30.9}$$

It is easy to see, on considering (30.4), that

$$m_\varkappa(\Omega, \tau) = \left(\frac{Q}{1+Q}\right)^{1/2} \chi(\Omega, \tau). \tag{30.10}$$

Now we calculate $I_\varkappa(\Omega,\tau)$. Substituting (30.7) into (30.9), we get

$$I_\varkappa(\Omega, \tau) = \Big\langle \Big[\left(\frac{Q}{1+Q}\right)^{1/2} \chi(\Omega, \tau) + \frac{1}{E_C}\left(\frac{Q}{1+Q}\right)^{1/2} \int_{-\infty}^{\infty} Z_C(t-\tau) X^*(t) \times$$
$$\times \exp(j\Omega t)\, dt\Big] \Big[\left(\frac{Q}{1+Q}\right)^{1/2} \chi^*(\Omega, \tau) + \frac{1}{E_C}\left(\frac{Q}{1+Q}\right)^{1/2} \times$$
$$\times \int_{-\infty}^{\infty} Z_C^*(t-\tau) X(t) \exp(j\Omega t)\, dt\Big]\Big\rangle.$$

Next, we multiply and statistically average the expression obtained, term by term:

$$I_\varkappa(\Omega, \tau) = \frac{Q}{1+Q}|\chi(\Omega, \tau)|^2 + \frac{Q}{E_C(1+Q)} \chi(\Omega, \tau) \int_{-\infty}^{\infty} \langle Z_C^*(t-\tau) X(t)\rangle \times$$
$$\times \exp(j\Omega t)\, dt + \frac{Q}{E_C(1+Q)} \chi^*(\Omega, \tau) \int_{-\infty}^{\infty} \langle Z_C(t-\tau) X^*(t)\rangle \times$$
$$\times \exp(j\Omega t)\, dt + \frac{Q}{E_C^2(1+Q)} \int_{-\infty}^{\infty}\int_{-\infty}^{\infty} Z_C(t'-\tau) Z_C^*(t''-\tau) \times$$
$$\times \langle X^*(t') X(t'')\rangle \exp[j\Omega(t'-t'')]\, dt'\, dt''. \tag{30.11}$$

In view of the fact that $Z_C(t)$ and $X(t)$ are not correlated, the second and third terms of (30.11) are zero. In examining the variance $d_\varkappa(\Omega,\tau)$, after substituting (30.10) and (30.11) into (30.8) and taking account of (30.6), we get

$$d_\varkappa(\Omega, \tau) = \frac{1}{E_C T_{\text{ef}}(1+Q)} \times$$

$$\times \int_{-\infty}^{\infty}\int_{-\infty}^{\infty} Z_C(t'-\tau)\, Z_C^*(t''-\tau)\, R_X(t'-t'')\exp\left[j\Omega\,(t'-t'')\right] dt'\,dt'', \tag{30.12}$$

where

$$R_X(t'-t'') = \frac{\langle X(t')\,X^*(t'')\rangle}{d_X}$$

is the autocorrelation coefficient for the random additive component of the echo signal $X(t)$. Substituting variables in (30.12) as $u' = t' - \tau$, $u'' = t'' - \tau$, we get the final expression for the variance $d_\varkappa(\Omega,\tau)$:

$$d_\varkappa(\Omega, \tau) = \frac{1}{E_C T_{\text{ef}}(1+Q)} \int_{-\infty}^{\infty}\int_{-\infty}^{\infty} Z_C(u')\, Z_C^*(u'') \times$$
$$\times R_X(u'-u'')\exp\left[j\Omega(u'-u'')\right] du'\,du''. \tag{30.13}$$

As an example, let us consider an emitted signal with sinusoidal occupation and a rectangular envelope. In this case

$$Z_C(t) = \begin{cases} 1, & t \in [0, T], \\ 0, & t \bar{\in} [0, T] \end{cases} \tag{30.14}$$

and, consequently,

$$E_C = T, \qquad T_{\text{ef}} = T. \tag{30.15}$$

Then on the basis of (30.13)-(30.15) we get

$$d_\varkappa(\Omega, \tau) = \frac{1}{T^2(1+Q)} \int_0^T\int_0^T R_X(u'-u'')\exp\left[j\Omega(u'-u'')\right] du'\,du''$$

$$= \frac{2}{T(1+Q)} \int_0^T \left(1 - \frac{u}{T}\right) R_X(u)\exp(j\Omega u)\, du. \tag{30.16}$$

Let us evaluate the average value $m_\varkappa(0,0)$ and the variance $d_\varkappa(0,0)$ at a point $\Omega = 0$, $\tau = 0$. In this case it follows from (30.10) that

$$m_\varkappa(0,0) = \left(\frac{Q}{1+Q}\right)^{1/2}, \tag{30.17}$$

and it follows from (30.16) that

$$d_\varkappa(0,0) = \frac{2}{T(1+Q)} \int_0^T \left(1 - \frac{u}{T}\right) R_X(u)\, du. \tag{30.18}$$

If $R_X(u)$ decreases rapidly enough in the interval $(0,T)$ as the correlation shift increases, we can assume the approximation

$$d_\varkappa(0,0) \approx \frac{2}{T(1+Q)} \int_0^T R_X(u)\, du = \frac{2}{\Delta F_X T(1+Q)}, \tag{30.19}$$

where

$$\Delta F_X = \frac{1}{\tau_X}, \qquad \tau_X = \int_0^T R_X(u)\, du,$$

in which case ΔF_X is the effective width of the spectrum of the echo signal's random component, and τ_X is the interval of its correlation. It should be kept in mind that relationship (30.19) is only valid when $\Delta F_X T \gg 1$.

On the basis of (30.17) and (30.19), we get

$$\gamma_\varkappa(0,0) \approx \left(\frac{2}{\Delta F_X T Q}\right)^{1/2}$$

for the coefficient of variation $\gamma_\varkappa(0,0)$.

The dependence of $m_\varkappa(0,0)$ and $d_\varkappa^{1/2}(0,0)$ on the size of of Q is shown in Figure 7. It follows from the figure that as the deterministic component of the echo signal increases -- that is, as parameter Q grows, the average value $m_\varkappa(0,0)$ increases and the mean square value $d_\varkappa^{1/2}(0,0)$ decreases. In addition, the greater the value of parameter $\Delta F_X T$, the smaller is the size of $d_\varkappa^{1/2}(0,0)$ for a fixed Q. This is natural, since averaging of the random component in the echo signal over signal duration T is more effective when the spectrum ΔF_X of this component is wider. It is interesting to note that in the case of emitted complex

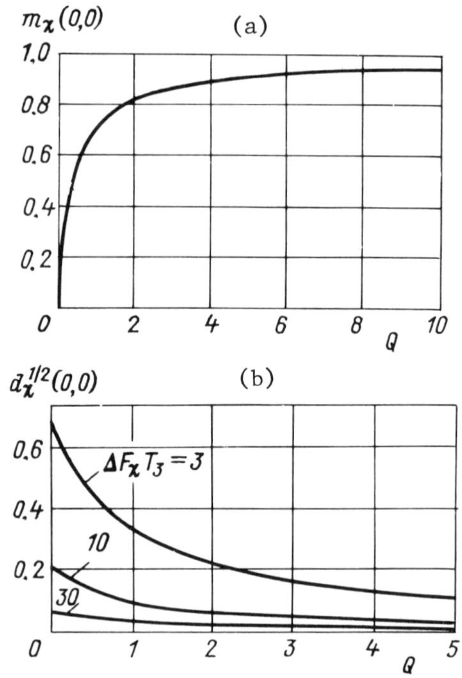

Figure 7. Dependence of the average value $m_\varkappa(0,0)$; (a), and mean square value $d_\varkappa^{1/2}(0,0)$; (b), on parameter Q for an echo signal with a random additive component.

signals parameter $\Delta F_X T$ is equivalent to their complexity factor. When an additive component $X(t)$ arises as a result of scattering, the variance $d_\varkappa(0,0)$ decreases as the complexity factor for the emitted signals grows, since the value of ΔF_X is proportional to the effective bandwidth of these signals. This last point means that conversion from simple emitted signals to complex signals increases (for the model being examined here at least) the cross correlation between the emitted signal and the echo signal.

§31. Echo Signal with Random Amplitude Modulation

In the case of an echo signal with random amplitude modulation we write

$$Z_S(t) = Z_C(t)\,[1 + m_A X(t)], \tag{31.1}$$

where $Z_C(t)$ is, as previously, the complex envelope of the emitted signal, m_A is the effective coefficient of the depth of random amplitude modulation, and $X(t)$ is a random stationary process with a mean of zero and variance of one:

$$m_X = \langle X(t) \rangle = 0, \qquad d_X = \langle | X(t) |^2 \rangle = 1. \tag{31.2}$$

Random amplitude modulation can arise in an echo signal because of the effect of irregularities in the water medium and its surface, which contribute multiplicative distortions to the echo signals, and because of specific features of the way acoustic waves reflect from moving objects of detection. Some of the causes for the appearance of random amplitude modulation in echo signals have been examined in §26.

The measure, $\hat{\varkappa}(\Omega,\tau)$, of the cross-correlation function in this case is defined by the relationship

$$\hat{\varkappa}(\Omega, \tau) \equiv \frac{1}{(E_C \langle E_s \rangle)^{1/2}} \int_{-\infty}^{\infty} Z_C(t-\tau)\, Z_C^*(t) \left[1 + m_A X^*(t)\right] \exp(j\Omega t)\, dt. \tag{31.3}$$

We find the average energy of the echo signal to be

$$\langle E_S \rangle = \left\langle \int_{-\infty}^{\infty} | Z_S(t) |^2 \, dt \right\rangle. \tag{31.4}$$

Substituting (31.1) into (3.14) we get

$$\langle E_S \rangle = \left\langle \int_{-\infty}^{\infty} | Z_C(t) |^2 [1 + m_A X(t)][1 + m_A X^*(t)]\, dt \right\rangle =$$

$$= E_C + m_A \int_{-\infty}^{\infty} | Z_C(t) |^2 \langle X(t) \rangle \, dt + m_A \int_{-\infty}^{\infty} | Z_C(t) |^2 \langle X^*(t) \rangle \, dt +$$

$$+ m_A^2 \int_{-\infty}^{\infty} | Z_C(t) |^2 \langle | X(t) |^2 \rangle \, dt.$$

In view of condition (31.2) the second and third terms of the last expression are equal to zero, and so we get

$$\langle E_S \rangle = E_C \left(1 + m_A^2\right). \tag{31.5}$$

Let us find the average value $m_\varkappa(\Omega,\tau)$. Considering (31.3) and (31.5) we have

$$m_\varkappa(\Omega, \tau) = \frac{1}{\left(1 + m_A^2\right)^{1/2}} \chi(\Omega, \tau) + \frac{m_A}{E_C \left(1 + m_A^2\right)^{1/2}} \int_{-\infty}^{\infty} Z_C(t-\tau) \langle X^*(t) \rangle \exp(j\Omega t)\, dt.$$

However, in view of (31.2) the second term of the expression obtained is equal to zero and, consequently,

$$m_{\varkappa}(\Omega, \tau) = \frac{1}{\left(1 + m_A^2\right)^{1/2}} \chi(\Omega, \tau). \tag{31.6}$$

Now let us define the expression for the variance $d_{\varkappa}(\Omega, \tau)$ of the quantity $\hat{\varkappa}(\Omega, \tau)$. First, we find an expression for

$$I_{\varkappa}(\Omega, \tau) = \langle |\hat{\varkappa}(\Omega, \tau)|^2 \rangle. \tag{31.7}$$

Substituting (31.3) into (31.7), we get

$$I_{\varkappa}(\Omega, \tau) = \frac{1}{E_C^2 \left(1 + m_A^2\right)} \int_{-\infty}^{\infty} \int_{-\infty}^{\infty} Z_C(t' - \tau) Z_C^*(t'' - \tau) \times$$

$$\times Z_C(t'') \langle [1 + m_A X^*(t)] [1 + m_A X(t')] \rangle \exp [j\Omega (t' - t'')] \, dt' \, dt''. \tag{31.8}$$

We next examine the product $\langle [1 + m_A X^*(t)] [1 + m_A X(t')] \rangle$ in the integrand of (31.8) in the light of (31.2). Then

$$\langle [1 + m_A X^*(t')] [1 + m_A X(t'')] \rangle = 1 + m_A \langle X(t') \rangle + m_A \langle X(t'') \rangle +$$

$$+ m_A^2 \langle X^*(t') X(t'') \rangle = 1 + m_A^2 R_X(t' - t''), \tag{31.9}$$

where $R_X(t' - t'') = \langle X^*(t')X(t'') \rangle$ is the coefficient of autocorrelation for the random modulating component of the echo signal. Now in accordance with (31.8) and (31.9), for $I_{\varkappa}(\Omega, \tau)$ we get

$$I_{\varkappa}(\Omega, \tau) = \frac{1}{1 + m_A^2} |\chi(\Omega, \tau)|^2 + \frac{m_A^2}{E_C \left(1 + m_A^2\right)^2} \int_{-\infty}^{\infty} \int_{-\infty}^{\infty} Z_C(t' - \tau) Z_C^*(t') \times$$

$$\times Z_C^*(t'' - \tau) Z_C(t'') R_X(t' - t'') \exp \left[j\Omega (t' - t'')\right] dt' \, dt''. \tag{31.10}$$

Going on to the variance $d_{\varkappa}(\Omega, \tau)$, using formulas (27.8), (31.6), and (31.10) we find that

$$d_{\varkappa}(\Omega, \tau) = \frac{m_A}{E_C \left(1 + m_A^2\right)} \int_{-\infty}^{\infty} \int_{-\infty}^{\infty} Z_C(t' - \tau) Z_C^*(t'') \times$$

$$\times Z_C^*(t'' - \tau) Z_C(t'') R_X(t' - t'') \exp \left[j\Omega (t' - t'')\right] dt' \, dt''. \tag{31.11}$$

At the point $\Omega = 0$, $\tau = 0$, for the average value $m_\varkappa(0,0)$ and variance $d_\varkappa(0,0)$ we have

$$m_\varkappa(0,0) = \frac{1}{(1+m_A^2)^{1/2}}; \tag{31.12}$$

$$d_\varkappa(0,0) = \frac{m_A^2}{E_C(1+m_A^2)^{1/2}} \int_{-\infty}^{\infty}\int_{-\infty}^{\infty} |Z_C(t')|^2 |Z_C(t'')|^2 R_X(t'-t'')\,dt'\,dt''. \tag{31.13}$$

Let us find the variance in the case where the emitted signal has a sinusoidal occupation, a rectangular envelope to the first order, and duration T. In this case $E_C = T$, and it follows from (31.13) that

$$d_\varkappa(0,0) = \frac{m_A^2}{T^2(1+m_A^2)} \int_0^T\int_0^T R_X(t'-t'')\,dt'\,dt'' = \frac{2m_A^2}{T(1+m_A^2)} \int_0^T \left(1-\frac{u}{T}\right) R_X(u)\,du. \tag{31.14}$$

As in the preceding case, we assume that

$$\tau_X = \int_0^T R_X(u)\,du, \qquad \Delta F_X = \frac{1}{\tau_X},$$

$$\int_0^T R_X(u)\,du \gg \frac{1}{T}\int_0^T uR_X(u)\,du, \tag{31.15}$$

that is, that correlation coefficient $R_X(u)$ decreases rather rapidly over the $(0,T)$ interval. Then, from (31.14) and (31.15) we have

$$d_\varkappa(0,0) \approx \frac{2m_A^2}{\Delta F_X T(1+m_A^2)}. \tag{31.16}$$

Comparing formulas (31.12) and (31.16) for $m_\varkappa(0,0)$ and $d_\varkappa(0,0)$ with formulas (30.17) and (30.19), we can see that they match when we set $m_A^2 = 1/Q$. This is natural, since parameter Q characterizes the relationship between the deterministic part of the echo signal and the random part, while m_A can be interpreted as the coefficient for the depth of random amplitude modulation. It is obvious that m_A^2 is the reciprocal of Q. We note, however, that there are differences, as well, in the behavior of the variance $d_\varkappa(\Omega,\tau)$ at $\Omega \neq 0$, $\tau \neq 0$ in the two cases examined in §30 and §31. This is indicated by the mismatch between the general formulas (30.13) and (31.11)

§32. Echo Signal with Random Phase Modulation

The complex envelope of an echo signal with random phase modulation has the form

$$Z_S(t) = Z_C(t)\exp[-j\Psi(t)], \tag{32.1}$$

where $\Psi(t)$ is the random instananeous phase of the echo signal [65]. Distortions in this type of echo signal can result from, for example, the effects of temporal fluctuations in the parameters of the water medium and the sea surface, as well as from the effects of specific features in the way acoustic waves reflect from the moving objects of research (in some cases) (see §26).

We will consider $\Psi(t)$ to be a stationary random process with a mean of zero, $m_\Psi = \langle\Psi(t)\rangle = 0$, and with variance $d_\Psi = \langle\Psi^2(t)\rangle$.

We write out the expression for the quantity $\hat{\varkappa}(\Omega,\tau)$ with a consideration for (32.1):

$$\hat{\varkappa}(\Omega,\tau) = \frac{1}{(E_C < E_S)^{1/2}}\int_{-\infty}^{\infty} Z_C(t-\tau)Z_C^*(t)\exp[j\Psi(t)]\exp(j\Omega t)\,dt. \tag{32.2}$$

We next define the average energy of the echo signal $\langle E_S\rangle$. It is easy to see that

$$\langle E_S\rangle \equiv \left\langle \int_{-\infty}^{\infty} |Z_S(t)|^2\,dt \right\rangle = E_C. \tag{32.3}$$

This means that random phase modulation does not change the echo signal's average energy. Now it follows from (32.2) and (32.3) that

$$\hat{\varkappa}(\Omega,\tau) = \frac{1}{E_C}\int_{-\infty}^{\infty} Z_C(t-\tau)Z_C^*(t)\exp[j\Psi(t)]\exp(j\Omega t)\,dt. \tag{32.4}$$

We now compute the average value $m_\varkappa(\Omega,\tau)$. Performing statistical averaging in (32.4), we get

$$m_\varkappa(\Omega,\tau) = \chi(\Omega,\tau)\langle\exp[j\Psi(t)]\rangle. \tag{32.5}$$

The average value $\langle\exp[j\Psi(t)]\rangle$ in formula (32.5) is nothing more than a one-dimensional characteristic function $\Theta_\Psi(\eta)$

of phase fluctuation $\Psi(t)$ with $\eta = -1$. Because $\Psi(t)$ is stationary it does not depend on current time t. And so we have

$$m_{\varkappa}(\Omega, \tau) = \chi(\Omega, \tau)\, \Theta_{\Psi}(-1). \tag{32.6}$$

Next, we determine the variance $d_{\varkappa}(\Omega,\tau)$. For this purpose we first find the characteristic

$$I_{\varkappa}(\Omega, \tau) = \langle |\hat{\varkappa}(\Omega, \tau)|^2 \rangle. \tag{32.7}$$

Substituting (32.4) into (32.7) we get

$$I_{\varkappa}(\Omega, \tau) = \frac{1}{E_C^2} \int_{-\infty}^{\infty}\int_{-\infty}^{\infty} Z_C(t'-\tau)\, Z_C^*(t''-\tau)\, Z_C(t'')\, Z_C^*(t'') \times \\ \times \langle \exp[j\Psi(t') - j\Psi(t'')] \rangle \exp[j\Omega(t'-t'')]\, dt'\, dt''. \tag{32.8}$$

It is not difficult to see what

$$\langle \exp[j\Psi(t') - j\Psi(t'')] \rangle = \Theta_{\Psi}(-1, 1; t''-t'), \tag{32.9}$$

where $\Theta_{\Psi}(\eta_1, \eta_2; t'' - t')$ is a two-dimensional characteristic function of phase $\Psi(t)$ at arguments $\eta_1 = -1$, $\eta_2 = 1$. Substituting (32.9) and (32.8) and going on to the variance $d_{\varkappa}(\Omega,\tau)$, considering (32.6) we get

$$d_{\varkappa}(\Omega, \tau) = \frac{1}{E_C^2} \int_{-\infty}^{\infty}\int_{-\infty}^{\infty} Z_C(t'-\tau)\, Z_C^*(t')\, Z_C^*(t''-\tau)\, Z_C(t'') \times \\ \times \Theta_{\Psi}(-1, 1; t''-t') \exp[j\Omega(t'-t'')]\, dt'\, dt'' - |\Theta_{\Psi}(-1)|^2\, |\chi(\Omega, \tau)|^2. \tag{32.10}$$

Let us examine a case in which $\Omega = 0$, $\tau = 0$, and the phase $\Psi(t)$ is a Gaussian, stationary random process. Then, considering that

$$\Theta_{\Psi}(-1) = \exp\left(-\frac{d_{\Psi}}{2}\right), \qquad d_{\Psi} \equiv \langle \Psi^2 \rangle, \tag{32.11}$$

for $m_{\varkappa}(0,0)$ we get

$$m_{\varkappa}(0, 0) = \exp\left(-\frac{d_{\Psi}}{2}\right).$$

In accordance with (32.10), the variance $d_\varkappa(0,0)$ is defined by the following expression:

$$d_\varkappa(0,0)=\frac{1}{E_C^2}\int_{-\infty}^{\infty}\int_{-\infty}^{\infty}|Z_C(t')|^2\,|Z_C(t'')|^2\,\Theta_\Psi(-1,1;t''-t')\,dt'\,dt''-|\Theta_\Psi(-1)|^2. \tag{32.12}$$

But for a Gaussian random process we have

$$\Theta_\Psi(-1,1;t''-t')=\exp\{-d_\Psi[1-R_\Psi(t''-t')]\}, \tag{32.13}$$

where $R_\Psi(t''-t')=\langle\Psi(t')\Psi(t'')\rangle/d_\Psi$ is the normalized coefficient of autocorrelation of the phase $\Psi(t)$.

Next, substituting (32.11) and (32.13) into (32.12), we get

$$d_\varkappa(0,0)=\frac{1}{E_C^2}\int_{-\infty}^{\infty}\int_{-\infty}^{\infty}|Z_C(t')|^2\,|Z_C(t'')|^2\times$$
$$\times\exp\{-d_\Psi[1-R_\Psi(t''-t')]\}\,dt'\,dt''-\exp(-d_\Psi). \tag{32.14}$$

Let us examine a case in which the exponent in the integrand of (32.14) is less than one -- that is, we can consider that $\exp(-v)\simeq 1-v$. Then, for $d_\varkappa(0,0)$, (32.14) can be written as the following approximate equality:

$$d_\varkappa(0,0)\approx\frac{d_\Psi}{E_C^2}\int_{-\infty}^{\infty}\int_{-\infty}^{\infty}|Z_C(t')|^2\,|Z_C(t'')|^2\,R_\Psi(t''-t')\,dt'\,dt''.$$

Let the emitted signal have a rectangular envelope to the first order, with duration T, and a sinusoidal occupation. In this case $E_C = T$. On the assumption that $R_\Psi(\tau)$ decreases rather rapidly over the interval $(0,T)$ -- that is, when $|R_\Psi(\tau)| \ll 1$, we have

$$d_\varkappa(0,0)\approx\frac{d_\Psi}{T^2}\int_0^T\int_0^T R_\Psi(t''-t')\,dt'\,dt''\approx\frac{d_\Psi}{\Delta F_\Psi T}, \tag{32.15}$$

where

$$\Delta F_\Psi=\frac{1}{\int_0^T R_\Psi(\tau)\,d\tau}$$

is the spectrum width of the function $\Psi(t)$.

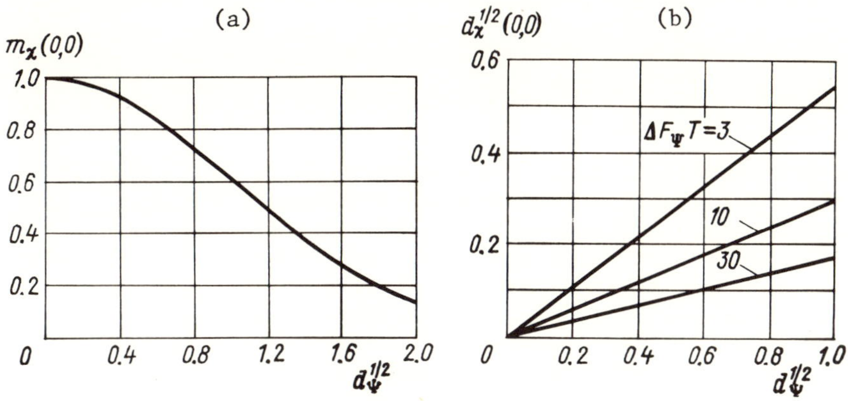

Figure 8. Dependence of the average value $m_\varkappa(0,0)$, (a), and mean square value $d_\varkappa^{1/2}(0,0)$, (b), on the mean square value $d_\Psi^{1/2}$ for an echo signal with random phase modulation.

Figure 8 illustrates the nature of the change of $m_\varkappa(0,0)$ and of $d_\varkappa^{1/2}(0,0)$ with the mean square value $d_\Psi^{1/2}$ of the fluctuations in phase $\Psi(t)$. It is evident from the graphs that the mathematical expectation $m_\varkappa(0,0)$ and the mean square $d_\varkappa^{1/2}(0,0)$ of the quantity $\hat{\varkappa}(0,0)$ depend in different ways on the mean square $d_\Psi^{1/2}$ of fluctuations in the instantaneous phase of the echo signal: The first parameter decreases while the second increases, as the fluctuation $d_\Psi^{1/2}$ increases. In addition to this, $d_\varkappa^{1/2}(0,0)$ decreases as the parameter $\Delta F_\Psi T$ grows. This is natural, since at large values of $\Delta F_\Psi T$, averaging phase fluctuations over the duration of the signal is more effective.

§33. Echo Signal with Random Frequency Modulation

Constructing a model of an echo signal with random frequency modulation [65], means representing its complex envelope in a way similar to that used in §32 -- that is,

$$Z_S(t) = Z_C(t)\exp[-j\Psi(t)],$$

with the difference that the random instantaneous phase $\Psi(t)$ must be defined by the relationship

$$\Psi(t) = \int_0^t \omega(t')\,dt', \tag{33.1}$$

where $\omega(t)$ is a random process characterizing changes in the echo signal's instantaneous frequency. The origin of random frequency modulation can involve fluctuations in the relative rate of movement velocity of the object of research (waviness on the sea surface can also be involved in the case when reflection from it is significant). In this case the relative instantaneous Doppler shift in frequency $\omega(t)$ of the echo signal can be defined as

$$\omega(t) = \frac{2\omega_0}{c}\,\vartheta(t), \tag{33.2}$$

where ω_0 is the emitted signal's carrier frequency, c is the rate of propagation of acoustic waves, and $\vartheta(t)$ are random fluctuations in the relative rate of movement of the reflecting object [65]. We will consider $\vartheta(t)$ to be a stationary random process. In accordance with (33.1) and (33.2) we have

$$\Psi(t) = \frac{2\omega_0}{c}\int_0^t \vartheta(t')\,dt'. \tag{33.3}$$

We define the measure $\hat{\varkappa}(\Omega,\tau)$ of the cross-correlation function here by

$$\hat{\varkappa}(\Omega,\tau) \equiv \frac{1}{(E_C\langle E_S\rangle)^{1/2}}\int_{-\infty}^{\infty} Z_C(t-\tau)\,Z_C^*(t)\exp[j\Psi(t)]\exp(j\Omega t)\,dt. \tag{33.4}$$

As in the preceding case (see §32), $\langle E_S\rangle = E_C$. We will examine emitted signals of the deterministic type. Therefore, to determine the characteristics of the measure $\hat{\varkappa}(\Omega,\tau)$ we must perform statistical averaging only with respect to the random phase $\Psi(t)$.

We get the mathematical expectation $m_\varkappa(\Omega,\tau)$ by statistical averaging of (33.4):

$$m_\varkappa(\Omega,\tau) = \frac{1}{E_C}\int_{-\infty}^{\infty} Z_C(t-\tau)\,Z_C^*(t)\,\Theta_\Psi(-1;\ t)\,\exp(j\Omega t)\,dt, \tag{33.5}$$

where

$$\Theta_{\Psi}(-1;\ t) = \langle \exp[j\Psi(t)] \rangle = \left\langle \exp\left[j\frac{2\omega_0}{c}\int_0^t \vartheta(t')\,dt'\right]\right\rangle$$

is a one-dimensional characteristic function (at $\eta = -1$) of the process $\Psi(t)$, which is nonstationary in this case.

To determine the variance $d_{\varkappa}(\Omega,\tau)$, we must first compute the function

$$I_{\varkappa}(\Omega,\ \tau) = \langle |\hat{\varkappa}(\Omega,\ \tau)|^2 \rangle. \tag{33.6}$$

Substituting (33.4) into (33.6) produces

$$I_{\varkappa}(\Omega,\ \tau) = \frac{1}{E_C^2}\int_{-\infty}^{\infty}\int_{-\infty}^{\infty} Z_C(t'-\tau)\,Z_C^*(t')\,Z_C^*(t''-\tau)\,Z_C(t'')\times$$
$$\times\,\Theta_{\Psi}(-1,\ 1;\ t',\ t'')\exp[j\Omega(t'-t'')]\,dt'\,dt'', \tag{33.7}$$

where

$$\Theta_{\Psi}(-1,\ 1;\ t',\ t'') = \langle \exp[j\Psi(t') - j\Psi(t'')] \rangle =$$
$$= \left\langle \exp\left[j\frac{2\omega_0}{c}\int_0^{t'}\vartheta(u)\,du - j\frac{2\omega_0}{c}\int_0^{t''}\vartheta(u)\,du\right]\right\rangle$$

is a two-dimensional characteristic function of the process $\Psi(t)$ with arguments $\eta_1 = 1$, $\eta_2 = 1$. Relationships (33.5) and (33.7) enable us to compute the variance $d_{\varkappa}(\Omega,\tau)$, which is defined by the expression

$$d_{\varkappa}(\Omega,\ \tau) = \frac{1}{E_C^2}\int_{-\infty}^{\infty}\int_{-\infty}^{\infty} Z_C(t'-\tau)\,Z_C^*(t')\,Z_C^*(t''-\tau)\,Z_C(t'')\times$$
$$\times\left[\Theta_{\Psi}(-1,\ 1;\ t',\ t'') - \Theta_{\Psi}(-1;\ t')\,\Theta_{\Psi}^*(-1;\ t'')\right]\exp[j\Omega(t'-t'')]\,dt'\,dt''. \tag{33.8}$$

Let us examine a particular case, in which $\vartheta(t)$ is a Gaussian stationary random process with a mean of zero, and variance

$$d_{\vartheta} = \langle \vartheta^2(t) \rangle, \tag{33.9}$$

and autocorrelation function

$$B_{\vartheta}(u) = \langle \vartheta(t+u)\,\vartheta(t) \rangle = d_{\vartheta}R_{\vartheta}(u), \tag{33.10}$$

where $R_\vartheta(u)$ is the (normalized) autocorrelation function of the process $\vartheta(t)$. In this case the characteristic functions $\Theta_\Psi(\eta,t)$ and $\Theta_\Psi(\eta_1,\eta_2;t',t'')$ have the following forms [12]:

$$\Theta_\Psi(\eta;\ t) = \exp\left[-\frac{d_\Psi(t)\,\eta^2}{2}\right], \tag{33.11}$$

$$\Theta_\Psi(\eta_1,\ \eta_2;\ t',\ t'') = \exp\left[-\frac{d_\Psi(t')\,\eta_1^2 + d_\Psi(t'')\,\eta_2^2 - 2B_\Psi(t',t'')\,\eta_1\eta_2}{2}\right]. \tag{33.12}$$

Using the relationships (33.3), (33.9), and (33.10), for $d_\Psi(t)$ and $B_\Psi(t',t'')$ we get

$$d_\Psi(t) = \frac{8\omega_0^2 d_\vartheta}{c^2}\int\limits_0^t (t'-u)\,R_\vartheta(u)\,du; \tag{33.13}$$

$$B_\Psi(t',t'') = \frac{4\omega_0^2 d_\vartheta}{c^2}\left[\int\limits_0^{t'} (t'-u)\,R_\vartheta(u)\,du + \int\limits_0^{t''} (t''-u)\,R_\vartheta(u)\,du - \right.$$
$$\left. - \int\limits_0^{|t'-t''|} (|t'-t''|-u)\,R_\vartheta(u)\,du\right] = \frac{d_\Psi(t') + d_\Psi(t'')}{2} -$$
$$- \frac{4\omega_0^2 d_\vartheta}{c^2}\int\limits_0^{|t'-t''|} (|t'-t''|-u)\,R_\vartheta(u)\,du. \tag{33.14}$$

We designate by

$$d_\omega = \frac{4\omega_0^2 d_\vartheta}{c^2} \tag{33.15}$$

the variance for the frequency excursion in the echo signal, which arises because of the random frequency modulation. Substituting (33.13) and (33.14) into (33.11) and (33.12) with a consideration for (33.15) gives (in this case we assume $\eta = 1$, $\eta_1 = -1$, $\eta_2 = 1$):

$$\Theta_\Psi(-1;\ t) = \exp\left[-d_\omega\int\limits_0^t (t-u)\,R_\vartheta(u)\,du\right], \tag{33.16}$$

$$\Theta_\Psi(-1,\ 1;\ t',\ t'') = \exp\left[-d_\omega\int\limits_0^{|t'-t''|} (|t'-t''|-u)\,R_\vartheta(u)\,du\right] \tag{33.17}$$

Next, we use relationships (33.16) and (33.17), substituting them into (33.5) and (33.8). Then, for the average value $m_\varkappa(\Omega,\tau)$ and the variance $d_\varkappa(\Omega,\tau)$ we get, correspondingly:

$$m_\varkappa(\Omega,\ \tau) = \frac{1}{E_C} \int_{-\infty}^{\infty} Z_C(t-\tau) Z_C^*(t) \exp\left[-d_\omega \int_0^t (t-u) R_\vartheta(u)\, du\right] \exp(j\Omega t)\, dt, \tag{33.18}$$

$$d_\varkappa(\Omega,\ \tau) = \frac{1}{E_C} \int_{-\infty}^{\infty} \int_{-\infty}^{\infty} Z_C(t'-\tau) Z_C^*(t') Z_C^*(t''-\tau) Z_C(t'') \times$$
$$\times \left\{\exp\left[-d_\omega \int_0^{|t'-t''|} (|t'-t''|-u) R_\vartheta(u)\, du\right] - \right.$$
$$- \exp\left[-2d_\omega \int_0^{t'} (t'-u) R_\vartheta(u)\, du - \right.$$
$$\left.\left. - 2d_\omega \int_0^{t''} (t''-u) R_\vartheta(u)\, du\right]\right\} \exp[j\Omega(t'-t'')]\, dt'\, dt''. \tag{33.19}$$

Let us evaluate the average value (33.18) at point $\Omega = 0$, $\tau = 0$, in the case when the inequality

$$d_\omega \int_0^t (t-u) R_\vartheta(u)\, du \ll 1$$

holds -- that is, when we can assume that

$$\exp\left[-d_\omega \int_0^t (t-u) R_\vartheta(u)\, du\right] \approx 1 - d_\omega \int_0^t (t-u) R_\vartheta(u)\, du. \tag{33.20}$$

We will assume that the emitted signal is a rectangular pulse of unit amplitude with duration T and a sinusoidal occupation, such that $E_C = T$. In this case it follows from (33.18) and (33.19) that

$$m_\varkappa(0,\ 0) \approx 1 - \frac{d_\omega}{T} \int_0^T \int_0^t (t-u) R_\vartheta(u)\, du\, dt. \tag{33.21}$$

We assume further that

$$R_\vartheta(u) = \exp(-|u|\Delta\omega_\vartheta), \tag{33.22}$$

where $\Delta\omega_\vartheta$ is the effective bandwidth of the spectrum of fluctuations in the velocity $\vartheta(t)$. Now we substitute (33.22) and (33.21), to get

$$m_\varkappa(0,\ 0) \approx 1 - \frac{3d_\omega T}{2\Delta\omega_\vartheta} - \frac{d_\omega}{\Delta\omega_\vartheta^3 T}[1 - \exp(-\Delta\omega_\vartheta T)]. \tag{33.23}$$

Designating the relative mean square value of spectrum expansion for the echo signal in expression (33.23) by A, viz.

$$A \equiv d_\omega^{1/2} T, \tag{33.24}$$

and the relative width of the spectrum of fluctuations in the rate of movement of the object of detection by B:

$$B \equiv \Delta\omega_\vartheta T. \tag{33.25}$$

then it follows from (33.23)-(33.25) that

$$m_\varkappa(0,\ 0) \approx 1 - \frac{3A^2}{2B} - \frac{A^2}{B^3}[1 - \exp(-B)].$$

If we now consider that $B \gg 1$, then for $m_\varkappa(0,0)$ we get

$$m_\varkappa(0,\ 0) \approx 1 - \frac{3A^2}{2B}. \tag{33.26}$$

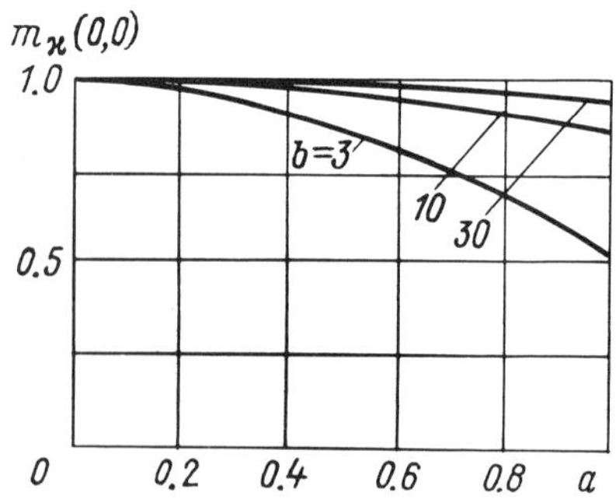

Figure 9. Dependence of the average value $m_\varkappa(0,0)$ on parameters $a = d_\omega^{1/2}T/2\pi$ and $b = \Delta\omega_\vartheta T/2\pi$, for an echo signal with random frequency modulation.

Figure 9 shows the relationships obtained in accordance with (33.26). Let us discuss the nature of the changes in $m_\varkappa(0,0)$ as changes occur in the parameters $a = A/2\pi$ and $b = B/2\pi$. As parameter a increases -- that is, as the fluctuation in the rate of movement of the object of detection increases, disruption of the echo signal intensifies, and $m_\varkappa(0,0)$ becomes smaller. But if the width of the spectrum of fluctuations increases -- that is, if parameter b grows, the smoothing effect on the echo signal over time T intensifies, causing $m_\varkappa(0,0)$ to increase.

§34. Model of an Echo Signal in the Form of a Sum of Elementary Signals

In response to the effects of irregularities in the water medium, reflection of acoustic waves from the medium's boundaries and, finally, in the event that objects of research have complex shapes, the echo signal at the point of observation may be found to be composed of several elementary echo signals. In this case the complex envelope $Z_S(t)$ takes the form:

$$Z_S(t) = \sum_{k=1}^{N} Z_{Sk}(t), \tag{34.1}$$

where N is the number of elementary signals, and the function $Z_{Sk}(t)$ characterizes the form of the kth elementary signal and depends in the general case on random parameters. The amplitude a_k, the moment of arrival t_k, and Doppler shift of frequency ω_k are examples of random parameters on which the elementary signal $Z_{Sk}(t)$ can depend.

Elementary signals can be interpreted physically as follows: They arise, for example, as a result of multiple-beam propagation of an acoustic wave, as a result of the presence of several reflecting surfaces on the object of detection, or in response to the effects of both of these causes.*

We recall that in complex form the emitted signal $Z(t)$ is written as:

$$Z(t) = Z_C(t) \exp(-j\omega_0 t), \tag{34.2}$$

---

*See §36 and footnote remarks. (D.M.)

where ω_0 is the central (angular) frequency of the signal's spectrum. By analogy with (32.2), we can write

$$Z_k(t) = a_k Z_C(t - t_k) \exp[-j(\omega_0 + \omega_k)(t - t_k)]. \tag{34.3}$$

for the kth elementary signal $Z_k(t)$.

We are interested in the complex envelope $Z_{Sk}(t)$ for the kth elementary signal $Z_k(t)$, which can be obtained from the following expression of the (34.2) class:

$$Z_k(t) = Z_{Sk}(t) \exp(j\omega_0 t). \tag{34.4}$$

Comparing (34.3) and (34.4), we find that

$$Z_{Sk}(t) = a_k Z_C(t - t_k) \exp[j\omega_0 t_k - j\omega_k(t - t_k)]. \tag{34.5}$$

Next, we go on to the representation of the complex envelope of the echo signal. In accordance with (34.1) and (34.5), the following expression is valid for the complex envelope $Z_S(t)$ of the echo signal:

$$Z_S(t) = \sum_{k=1}^{N} a_k Z_C(t - t_k) \exp[j\omega_0 t_k - j\omega_k(t - t_k)]. \tag{34.6}$$

Now we define some characteristics of the echo signal pertaining to the model under examination. We find the quantity $\hat{\varkappa}(\Omega,\tau)$ for the coefficient of cross correlation between the emitted signal and the echo signal to be:

$$\hat{\varkappa}(\Omega, \tau) = \frac{1}{(E_C \langle E_S \rangle)^{1/2}} \sum_{k=1}^{N} a_k \exp[-j(\omega_0 + \omega_k) t_k] \times$$
$$\times \int_{-\infty}^{\infty} Z_C(t - \tau) Z_C^*(t - t_k) \exp[-j(\Omega - \omega_k) t]\, dt. \tag{34.7}$$

Assuming that $\omega_k \ll \omega_0, \Omega \ll \omega_0$, and converting to the ambiguity function in (34.7), we have

$$\int_{-\infty}^{\infty} Z_C(t - \tau) Z_C^*(t - t_k) \exp[-j(\Omega - \omega_k) t]\, dt =$$

$$= \int_{-\infty}^{\infty} Z_C\left[x-(\tau-t_k)\right] Z_C^*(x) \exp\left[-j(\Omega-\omega_k)(x-t_k)\right] dx =$$
$$= E_C \exp[-j(\Omega-\omega_k)t_k]\,\chi(\Omega-\omega_k,\ t-t_k). \qquad (34.8)$$

Then, obviously, on the basis of (34.7) and (34.8) for $\hat{\varkappa}(\Omega,\tau)$ we get

$$\hat{\varkappa}(\Omega,\ \tau) = \left(\frac{E_C}{\langle E_S\rangle}\right)^{1/2} \sum_{k=1}^{N} a_k \exp(-j\omega_0 t_k)\,\chi(\Omega-\omega_k,\ \tau-t_k). \qquad (34.9)$$

We will subsequently assume that the k ($\neq 1$) random parameters a_k, t_k, and ω_k are statistically independent and that their distribution laws do not depend on the parameter k of the elementary echo signal.

Let us find the average energy $\langle E_S\rangle$ of the echo signal. In accordance with (34.6) we have

$$\langle E_S\rangle = \left\langle \sum_{k=1}^{N}\sum_{l=1}^{N} a_k a_l \int_{-\infty}^{\infty} Z_C(t-t_k)\,Z_C^*(t-t_l)\,dt \right\rangle.$$

Remembering our statement on the independence of the statistical properties of the random parameters describing elementary signals, for $\langle E_S\rangle$ we get

$$\langle E_S\rangle = E_C[\,N\langle a^2\rangle + (N^2-N)\langle a\rangle^2] = NE_C[\langle a^2\rangle - \langle a\rangle^2 + N\langle a\rangle^2].$$

Substituting (34.10) into (34.9), we find that

$$\hat{\varkappa}(\Omega,\ \tau) = \frac{1}{\{N[\langle a^2\rangle - \langle a\rangle^2 + N\langle a\rangle^2]\}^{1/2}} \sum_{k=1}^{N} a_k \exp(-j\omega_0 t_k)\,\chi(\Omega-\omega_k,\ \tau-t_k). \qquad (34.11)$$

Because the band of the emitted signals is narrow, it is natural to consider that the factor $\exp(-j\omega_0 t_k)$ in expression (34.11) changes faster with t_k than does the ambiguity function $\chi(\Omega-\omega_k, \tau - t_k)$. We will write

$$t_k = \tau_k + \Delta t_k, \qquad \omega_k = \Omega_k - \Delta\omega_k,$$

where $\tau_k = \langle t_k \rangle$, $\Omega_k = \langle \omega_k \rangle$. Then, from (34.11) we arrive at the following relationship

$$\hat{\varkappa}(\Omega, \tau) = \frac{1}{\{N[\langle a^2 \rangle - \langle a \rangle^2 + N\langle a \rangle^2]\}^{1/2}} \times$$
$$\times \sum_{k=1}^{N} a_k \exp(-j\omega_0 \tau_k) \exp(-j\omega_0 \Delta t_k) \chi(\Omega - \Omega_k - \Delta\omega_k, \tau - \tau_k - \Delta t_k). \tag{34.12}$$

Let us find the average value $m_\varkappa(\Omega, \tau)$. Averaging (34.12) under the assumption made above that the distributions of the random parameters do not depend on the k-number and that the parameters themselves are not statistically related, we get

$$m_\varkappa(\Omega, \tau) = \frac{1}{\{N[\langle a^2 \rangle - \langle a \rangle^2 + N\langle a \rangle^2]\}^{1/2}} \times$$
$$\times \Theta_{\Delta t}(\omega_0) \sum_{k=1}^{N} \exp(-j\omega_0 \tau_k) \langle \chi(\Omega - \Omega_k - \Delta\omega, \tau - \tau_k) \rangle, \tag{34.13}$$

where $\Theta_{\Delta t}(\omega_0) = \langle \exp(-j\omega_0 \Delta t) \rangle$ is the characteristic function of the distribution of random fluctuations Δt_k at the times elementary signals arrive.

Let us examine a particular case when $\tau_k = 0$, $\Omega_k = 0$, $N \gg 1$, which corresponds to identical average times of arrival for a rather large number of elementary echo signals with identical average shifts. In this case, from (34.13) we have

$$m_\varkappa(\Omega, \tau) \approx \Theta_{\Delta t}(\omega_0) \langle \chi(\Omega - \Delta\omega, \tau) \rangle, \tag{34.14}$$

where

$$\langle \chi(\Omega - \Delta\omega, \tau) \rangle = \int_{-\infty}^{\infty} W_{\Delta\omega}(\Delta\omega) \chi(\Omega + \Delta\omega, \tau) \, d\Delta\omega, \tag{34.15}$$

and $W_{\Delta\omega}(\Delta\omega)$ is the probability density of the random frequency shifts $\Delta\omega$.

It follows from (34.14), in particular, that the greater the fluctuation of random variables Δt and $\Delta\omega$, the smaller is the mathematical expectation $m_{\varkappa}(\Omega,\tau)$. Obviously

$$\max m_{\varkappa}(\Omega, \tau) = \chi(\Omega, \tau),$$

which corresponds to the condition $\Delta t = 0$, $\Delta\omega = 0$ -- that is, to the absence of fluctuations in the parameters of the elementary signals, and to their coherent summation.

CHAPTER VI

MARINE REVERBERATION

§35. Classification of Reverberatory Signals

When emitted signals propagate they are partially scattered by various irregularities in the water medium and by unevenness in its boundaries. Such scattering could be caused by air bubbles, objects of biological origin (fish, crustaceans, microorganisms), temperature irregularities, unevenness of the sea surface, unevenness of the bottom, irregularities in the bottom's composition, and so on.

We define marine reverberation as a random process (actually, a random field in space as well as time) observed at the point of reception after a sonar signal is injected into the water medium. It is a random process which describes the change of the total scattered signal with time.

Studying the probability characteristics of reverberation is interesting from two standpoints. One aspect is studying the properties of reverberation as sonar interference -- that is, as a factor hindering detection of echo signals and the measurement of their parameters. The second aspect is studying reverberation as a phenomenon with which we could investigate and predict the acoustic properties of the water medium, its boundaries, and other objects of research. The methods and levels of probabilistic description in these two cases can differ from one another. We will examine probability models of marine reverberation below, developing them primarily for solving inverse probability problems -- that is, studying the properties of the water medium, its boundaries, and its irregularities on the basis of the chacteristics of reverberation.

Three types of reverberation exist -- three-dimensional, reverberation from a layer, and from a boundary.

They basically differ with respect to the geometric distribution of scattering irregularities in the water medium (a volume, a layer, a boundary).

Table 3 provides a classification of reverberatory signals and indicates the scattering objects. It stands to reason that combinations of different types of reverberatory signals can be observed at the point of reception in sonar studies.

This classification is useful in examining the characteristics of reverberation, and it enables us to develop probability models of reverberation.

§36. Discrete Canonical Model of Reverberation

We use the so-called discrete model of the scattering of acoustic waves in the water medium to study the probability characteristics of marine reverberation. In this case we rely on the discrete canonical model of reverberation is represented as a sum of elementary scattered signals [14,17,33,35,39,52,53,58].

Table 3

Type of reverberation	Distribution of scatterers in water medium	Scattering objects
Three-dimensional	Throughout the entire volume embraced by the acoustic antenna beam patterns and the signal emitted into the water medium	Biological accumulations, air bubbles, temperature irregularities
Reverberation from a layer	In a layer embraced by acoustic antenna beam patterns	Air bubbles in subsurface layer, accumulations of biological objects
Boundary	Along a boundary	Unevenness on sea bottom or sea surface

We will examine the reverberation model as a temporal random process $F(t)$ under the following assumptions:

(a) the scattering irregularities are discrete;

(b) secondary scattering is absent;

(c) the medium is homogeneous with respect to refractive distortions within it and the effects of reflecting boundaries (i.e., $\nabla c = 0$);

(d) the acoustic emitter and receiver are omnidirectional.*

Under these assumptions the discrete canonical model of the reverberation process $F(t)$ is described by

$$F(t) = \sum_{k=1}^{N} f_k(t), \tag{36.1}$$

where N is the number of elementary scattered signals creating reverberation at the point of reception at time t, and $f_k(t)$ is a function describing the form of the kth elementary signal. The $f_k(t)$ can be expanded as:

$$f_k(t) = \alpha_k \varphi(t_k) C(t - t_k, \varepsilon_k), \tag{36.2}$$

where α_k is a random variable characterizing the scattering properties of the kth scatterer, $\varphi(t_k)$ is a certain function describing the diminishing of signals as they propagate through the water medium, t_k is the time at which the kth elementary scattered signal arrives at the point of reception, $C(t)$ is a function describing the emitted signal, and the ε_k are random parameters defining the form of the elementary scattered signals. Examples of such random parameters are the Doppler frequency shift, stemming from move-

---

*This important, simplifying assumption is in force throughout the remaining chapters. Consequently, the effects of these transmitting and receiving apertures appear as simple scale factors in the waveform amplitudes. In the preceding Chapters IV, V, however, it is implicitly assumed that the (complex) emitted and echo signals, Z_c, Z_s, include the effects of any filtering action by the transmitting and receiving apertures (arrays). (D.M.)

ment of the scatterers, the direction of the wave front of the elementary signal which depends on the location of the signal source, and so on.

In the case where the form of elementary signals $f_k(t)$ is constant -- that is, $C(t - t_k, \varepsilon_k) = C(t - t_k)$, a simpler expression is valid for the $f_k(t)$:

$$f_k(t) = \alpha_k \varphi(t_k) C(t - t_k). \tag{36.3}$$

Thus, we assume that reverberation at the point of reception is a sum of elementary signals scattered by discrete irregularities in the water medium, and we will use its model, which according to (36.1)-(36.3) has the form:

$$F(t) = \sum_{k=1}^{N} \alpha_k \varphi(t_k) C(t - t_k, \varepsilon_k), \tag{36.4}$$

or in the simpler case,

$$F(t) = \sum_{k=1}^{N} \alpha_k \varphi(t_k) C(t - t_k). \tag{36.5}$$

§37. Complex Representations

Let us examine a complex representation of reverberation using model (36.5) as its basis.

We represent the emitted signal by the real function

$$C(t) = A(t) \cos [\omega_0 t + \Phi(t)], \tag{37.1}$$

as had been done previously (see §21), which, in complex form, is

$$Z(t) = C(t) + j\widetilde{C}(t), \tag{37.2}$$

or

$$Z(t) = Z_C(t) \exp(-j\omega_0 t), \tag{37.3}$$

where $C(t)$ and $\tilde{C}(t)$ are Hilbert conjugated processes, and $Z_C(t)$ is the complex envelope of the emitted signal.

We substitute (37.1) into (36.5). In this case the real function $F(t)$ is written as:

$$F(t) = \sum_{k=1}^{N} \alpha_k \varphi(t_k) A(t - t_k) \cos[\omega_0 (t - t_k) + \Phi(t - t_k)]. \tag{37.3}$$

Now let

$$Y(t) = F(t) + j\tilde{F}(t) \tag{37.4}$$

be a complex random function corresponding to a real function $F(t)$, in which case $\tilde{F}(t)$ is a Hilbert transformation of $F(t)$. We represent function $Y(t)$ as

$$Y(t) = Y_F(t) \exp(-j\omega_0 t), \tag{37.5}$$

where function $Y_F(t)$ is analogous to $Z_C(t)$ in (37.2) and can be called the complex envelope of marine reverberation. On the basis of (37.3)-(37.5) we get

$$Y(t) = \sum_{k=1}^{N} \alpha_k \varphi(t_k) A(t - t_k) \exp[-j\omega_0 (t - t_k) - j\Phi(t - t_k)].$$

Next we recall that the complex envelope of an emitted signal had been represented as

$$Z_C(t) = A(t) \exp[-j\Phi(t)].$$

Then we can define the complex envelope of the reverberation $Y_F(t)$ as follows:

$$Y_F(t) = \sum_{k=1}^{N} \alpha_k \varphi(t_k) Z_C(t - t_k) \exp(-j\omega_0 t_k), \tag{37.6}$$

or in more compact form,

$$Y_F(t) = \sum_{k=1}^{N} y_{Fk}(t), \tag{37.7}$$

where

$$y_{Fk}(t) = \alpha_k \varphi(t_k) Z_C(t - t_k) \exp(-j\omega_0 t_k) \quad (37.8)$$

is the complex envelope of the kth elementary scattered signal.

§38. Characteristics of the Random Parameters of Elementary Scattered Signals

We must have information on the probability distributions of random parameters N, α_k, and t_k if we are to obtain various probability characteristics of the reverberation. If we assume that scatterers are distributed independently in the water medium and do not affect one another from the standpoint of scattering properties, then these random parameters can be considered statistically independent. In addition there are sufficient grounds for making one more assumption, namely, that the probability distribution laws for the random parameters do not depend on the k-number. All of this enables us to perform statistical averaging with respect to parameters N, α_k, and t_k separately.

We note that such a scattering model had been examined in most works (see, for example, [14,33,35,39])* in which the authors limited themselves to a phenomenological discrete canonical model of marine reverberation. In a number of cases the properties of the random parameters formulated above have been substantiated from a physical standpoint.

§39. Average Value of Reverberation

Let us find the average value of the complex envelope of reverberation

$$m_F(t) = \langle Y_F(t) \rangle.$$

Taking account of (37.7) we can write

$$m_F(t) = \langle \sum_{k=1}^{N} y_{Fk}(t) \rangle.$$

---

*For example, the so-called FOM theory [cf. Preface], also, [53,58,62]. (D.M.)

Let $P(N,t)$ be the distribution law for N elementary scattered signals. On the basis of the assumptions formulated in §38, we have

$$m_F(t) = \sum_{N=0}^{\infty} NP(N, t)\langle y_F(t)\rangle, \tag{39.1}$$

in which case indices k are dropped when the average is taken in (39.1) because it is assumed that the probability characteristics of the elementary signals do not depend the k-number, so that $\langle y_{Fk}(t)\rangle = \langle y_F(t)\rangle$.

We introduce a function $n(t)$ to define the average number of elementary scattered signals arriving at the point of reception within a unit of time. Then

$$m_N(t) = \sum_{N=0}^{\infty} NP(N, t) = n(t)T_{\text{ef}}, \tag{39.2}$$

where $m_N(t)$ is the average number of elementary oscillations generating reverberation at time t, and T_{ef} is the effective duration of the emitted signal. We will subsequently consider that $t_k = t'$, that function $\varphi(t')$ changes insignificantly within the interval $(t-T_{\text{ef}}/2, t+T_{\text{ef}}/2)$, and in correspondence with this we assume

$$\varphi(t') = \varphi(t). \tag{39.3}$$

Now let us examine the average value

$$\langle y_F(t)\rangle = \langle \alpha\rangle \varphi(t)\langle Z_C(t-t')\exp(-j\omega_0 t')\rangle, \tag{39.4}$$

where $\langle\alpha\rangle$ is the average of random variables α.

Next, substituting (39.2) and (39.4) into (39.1) for the average value $m_F(t)$ we get

$$m_F(t) = \langle\alpha\rangle n(t)\varphi(t)T_{\text{ef}}\langle Z_C(t-t')\exp(-j\omega_0 t')\rangle. \tag{39.5}$$

We now add the probability distribution of random variables t' to our discussion. We can assume on sufficient grounds (see, for example [15,17,33]) that the probability density $W_t(t')$ is uniform in the interval $(t-T_{ef}/2, t+T_{ef}/2)$ -- that is,

$$W_t(t') = \begin{cases} 1/T_{ef}, & t' \in [t \pm T_{ef}/2]; \\ 0, & t' \overline{\in} [t \pm T_{ef}/2]. \end{cases} \tag{39.6}$$

Then

$$\begin{aligned} \langle Z_C(t-t')\exp(-j\omega_0 t')\rangle &= \int_{-\infty}^{\infty} W_t(t') Z_C(t-t') \exp(-j\omega_0 t')\, dt' \\ &= \frac{1}{T_{ef}} \int_{t-T_{ef}/2}^{t+T_{ef}/2} Z_C(t-t') \exp(-j\omega_0 t')\, dt' \\ &= \frac{\exp(-j\omega_0 t)}{T_{ef}} \int_{-T_{ef}/2}^{T_{ef}/2} Z_C(x) \exp(j\omega_0 x)\, dx. \end{aligned} \tag{39.7}$$

Next, we consider that $g_C(\omega)$ is the frequency spectrum for the complex envelope of signal $Z_C(t)$:

$$g_C(\omega) = \int_{-\infty}^{\infty} Z_C(t) \exp(-j\omega t)\, dt. \tag{39.8}$$

Without noticeable error, ignoring edge effects, we can expand the integration interval $(-T_{ef}/2, T_{ef}/2)$ in expression (39.7) to $(-\infty, \infty)$. Then, substituting (39.7) into (39.5) and considering (39.8), we get

$$m_F(t) \approx \langle\alpha\rangle n(t) \varphi(t) \int_{-\infty}^{\infty} Z_C(x) \exp(j\omega_0 x)\, dx = \langle\alpha\rangle n(t) \varphi(t) g_C(-\omega_0).$$

However, in view of the fact that the emitted signal has a narrow band

$$g_C(-\omega_0) \approx 0, \tag{39.9}$$

we can say that

$$m_F(t) \approx 0. \tag{39.10}$$

It should be kept in mind [14,16,35,37] that in cases when emitted signals have wide bands, or when so-called coherent scattering occurs at the reception point, the average value

$m_F(t)$ is not equal to zero. Some characteristics of reverberation applicable to such cases are examined in §§44,46.

§40. Autocorrelation Function for the Complex Envelope of Reverberation

Let us examine the reverberation process $Y_F(t)$ at two points in time: $t - \tau/2$ and $t + \tau/2$:

$$\left.\begin{aligned} Y_F\left(t-\frac{\tau}{2}\right) &= \sum_{k=1}^{N} y_{Fk}\left(t-\frac{\tau}{2}\right); \\ Y_F\left(t+\frac{\tau}{2}\right) &= \sum_{k=1}^{N} y_{Fk}\left(t+\frac{\tau}{2}\right). \end{aligned}\right\} \tag{40.1}$$

By definition, the current-t autocorrelation function $B_F(t,\tau)$ of the complex envelope of reverberation $Y_F(t)$ equals

$$B_F(t,\tau) = \left\langle Y_F\left(t-\frac{\tau}{2}\right) Y_F^*\left(t+\frac{\tau}{2}\right)\right\rangle. \tag{40.2}$$

Substituting (40.1) into (40.2) we get

$$B_F(t,\tau) = \left\langle \sum_{k=1}^{N} \sum_{l=1}^{N} y_{Fk}\left(t-\frac{\tau}{2}\right) y_{kF}^*\left(t+\frac{\tau}{2}\right)\right\rangle. \tag{40.3}$$

We proceed as in §39 and introduce the distributin $P(N,t)$. We separate out two terms in the double summation of (40.3) -- the first for $k = l$, and the second for $k \neq l$. Then we have

$$B_F(t,\tau) = \sum_{N=0}^{\infty} NP(N,t)\left\langle y_F\left(t-\frac{\tau}{2}\right) y_F^*\left(t+\frac{\tau}{2}\right)\right\rangle +$$

$$+ \sum_{N=0}^{\infty} P(N,t) \sum_{\substack{k=1 \\ k\neq l}}^{N} \sum_{l=1}^{N} \left\langle y_{Fk}\left(t-\frac{\tau}{2}\right) y_{Fl}^*\left(t+\frac{\tau}{2}\right)\right\rangle. \tag{40.4}$$

It is easy to see that on the basis of assumptions made in §38 the second term of (40.4) equals zero. Therefore, considering (39.10) and (40.2) we find that

$$B_F(t, \tau) = n(t)\, T_{\text{ef}} \left\langle y_F\left(t - \frac{\tau}{2}\right) y_F^*\left(t + \frac{\tau}{2}\right)\right\rangle. \qquad (40.5)$$

We make use of expression (37.8) and take account of the statistical independence of the variables $\alpha_k = \alpha$, $t_k = t'$ and of condition (39.3). Then we get

$$\left\langle y_F\left(t - \frac{\tau}{2}\right) y_F^*\left(t + \frac{\tau}{2}\right)\right\rangle =$$

$$\langle\alpha^2\rangle\, \varphi^2(t) \left\langle Z_C\left(t - \frac{\tau}{2} - t'\right) Z_C^*\left(t + \frac{\tau}{2} - t'\right)\right\rangle. \qquad (40.6)$$

Assuming as previously that the distribution of the random variable t' is uniform and of the type (39.6), in accordance with (40.5) and (40.6), for $B_F(t,\tau)$ we have

$$B_F(t, \tau) = \langle\alpha^2\rangle\, n(t)\, \varphi^2(t) \int_{t-T_{\text{ef}}/2}^{t+T_{\text{ef}}/2} Z_C\left(t - \frac{\tau}{2} - t'\right) Z_C^*\left(t + \frac{\tau}{2} - t'\right) dt'. \qquad (40.7)$$

We make the substitution $x = t - t'$ in the integrand of (40.7) and, ignoring edge effects, we extend the domain of integration to $(-\infty, \infty)$. Then from (40.7) we arrive at the following final expression for the autocorrelation function of the complex envelope of reverberation:

$$B_F(t, \tau) = \langle\alpha^2\rangle\, n(t)\, \varphi^2(t) \int_{-\infty}^{\infty} Z_C\left(x - \frac{\tau}{2}\right) Z_C^*\left(x + \frac{\tau}{2}\right) dx. \qquad (40.8)$$

Next, let us find the variance of the complex envelope of reverberation $d_F(t)$. Inasmuch as $d_F(t) = B_F(t,0)$, it follows from (40.8) that

$$d_F(t) = \langle\alpha^2\rangle\, n(t)\, \varphi^2(t)\, E_C, \qquad (40.9)$$

where

$$E_C = \int_{-\infty}^{\infty} |Z_C(x)|^2\, dx$$

is the energy of the emitted signal.

The (normalized) coefficient of autocorrelation for the complex envelope of reverberation

$$R_F(\tau) \equiv \frac{B_F(t, \tau)}{d_F(t)}$$

is defined on the basis of (40.8) and (40.9) by the expression

$$R_F(\tau) = \frac{1}{E_C} \int_{-\infty}^{\infty} Z_C\left(x - \frac{\tau}{2}\right) Z_C^*\left(x + \frac{\tau}{2}\right) dx. \qquad (40.10)$$

It is evident from the relationships obtained that reverberation is a nonstationary random process. However, given the assumptions made, we can see that it relates to a class of processes that can be reduced to a stationary state, as is indicated by the independence of the coefficient of autocorrelation (40.10) from current time (t).

§41. Energy Spectrum of the Complex Envelope of Reverberation

Expression (40.8) for the correlation function $B_F(t,\tau)$ enables us to determine the current-t energy spectrum $G_F(t,\omega)$ of reverberation, which is associated with $B_F(t,\tau)$ by the relationship

$$G_F(t, \omega) = \int_{-\infty}^{\infty} B_F(t, \tau) \exp(-j\omega\tau)\, d\tau. \qquad (41.1)$$

Substituting (40.8) into (41.1), we find that

$$G_F(t, \omega) = \langle \alpha^2 \rangle n(t) \varphi^2(t) \int_{-\infty}^{\infty} \int_{-\infty}^{\infty} Z_C\left(x - \frac{\tau}{2}\right) Z_C^*\left(x + \frac{\tau}{2}\right) \exp(-j\omega\tau)\, dx\, d\tau. \qquad (41.2)$$

Let us examine the double integral in expression (41.2):

$$I(t, \omega) = \int_{-\infty}^{\infty} \int_{-\infty}^{\infty} Z_C\left(x - \frac{\tau}{2}\right) Z_C^*\left(x + \frac{\tau}{2}\right) \exp(-j\omega\tau)\, dx\, d\tau. \qquad (41.3)$$

We replace the functions $Z_C(x - \tau/2)$ and $Z_C^*(x + \tau/2)$ by their frequency spectra:

$$\left.\begin{aligned} Z_C\left(x-\frac{\tau}{2}\right) &= \frac{1}{2\pi}\int\limits_{-\infty}^{\infty} g_C(\omega)\exp\left[j\omega\left(x-\frac{\tau}{2}\right)\right]d\omega; \\ Z_C^*\left(x+\frac{\tau}{2}\right) &= \frac{1}{2\pi}\int\limits_{-\infty}^{\infty} g_C^*(\omega)\exp\left[-j\omega\left(x+\frac{\tau}{2}\right)\right]d\omega. \end{aligned}\right\} \qquad (41.4)$$

Substituting (41.4) into (41.3) we get

$$I(t,\omega) = \frac{1}{2\pi}\int\limits_{-\infty}^{\infty}\int\limits_{-\infty}^{\infty}\int\limits_{-\infty}^{\infty}\int\limits_{-\infty}^{\infty} g_C(\omega')\,g_C^*(\omega'')\times$$

$$\times\exp\left[j\omega'\left(x-\frac{\tau}{2}\right)-j\omega''\left(x+\frac{\tau}{2}\right)-j\omega\tau\right]d\omega'\,d\omega''\,dx\,d\tau. \qquad (41.5)$$

We integrate (41.5) with respect to the variable x and consider that

$$\delta(y) = \frac{1}{2\pi}\int\limits_{-\infty}^{\infty}\exp(\pm jyx)\,dx \qquad (41.6)$$

is a delta-function. We next substitute (41.6) into (41.5) and take account of the following property of a delta-function:

$$\int\limits_{-\infty}^{\infty} P(y)\,\delta(y-y_0)\,dy = P(y_0). \qquad (41.7)$$

Then we get

$$I(t,\omega) = \frac{1}{2\pi}\int\limits_{-\infty}^{\infty}\int\limits_{-\infty}^{\infty}\int\limits_{-\infty}^{\infty} g_C(\omega')\,g_C^*(\omega'')\exp\left[-j\left(\frac{\omega'+\omega''}{2}+\omega\right)\tau\right]\times$$

$$\times\,\delta(\omega'-\omega'')\,d\omega'\,d\omega''\,d\tau = \frac{1}{2\pi}\int\limits_{-\infty}^{\infty}\int\limits_{-\infty}^{\infty}|g_C(\omega'')|^2\exp[-j(\omega''+$$

$$+\,\omega)\tau]\,d\omega''\,d\tau = \int\limits_{-\infty}^{\infty}|g_C(\omega'')|^2\,\delta(\omega''+\omega)\,d\omega = |g_C(-\omega)|^2 = |g_C(\omega)|^2. \qquad (41.8)$$

This means that in accordance with (41.2) and (41.8) the current-t (e.g., instantaneous) energy spectrum of reverberation is defined by the relationship

$$G_F(t, \omega) = \langle \alpha^2 \rangle n(t) | g_C(\omega) |^2. \tag{41.9}$$

We can see that the energy spectrum of marine reverberation matches in form the square of the modulus of the spectrum for the emitted signal's complex envelope.*

§42. Evaluation of the Frequency-Time Cross-Correlation Function for the Complex Envelopes of the Emitted Signal and Reverberation

The short-term, normalized function describing cross correlation between the emitted signal and the reverberation is an important characteristic of reverberatory signals. Designating this function by $\hat{r}(\Omega, t)$, we define it similarly to the way this had been done in the analysis of echo signals (see §27), using the characteristic of signal similarity:

$$\hat{r}(\Omega, t) \equiv \frac{1}{[E_C d_F(t) T_{ef}]^{1/2}} \int_{-\infty}^{\infty} Z_C(t' - t) Y_F^*(t') \exp(j\Omega t')\, dt', \tag{42.1}$$

where

$$E_C = \int_{-\infty}^{\infty} | Z_C(t) |^2\, dt$$

is the energy of the emitted signal, the complex envelope of which is described by function $Z_C(t)$, T_{ef} is the effective duration of the emitted signal, and

$$d_F(t) = \langle | Y_F(t) |^2 \rangle = \langle \alpha^2 \rangle n(t) \varphi^2(t) E_C \tag{42.2}$$

is the variance of the reverberation at time t.

We make use of the representations of $Y_F(t)$ given above:

$$Y_F(t) = \sum_{k=1}^{N} y_{Fk}(t). \tag{42.3}$$

---

*This is a form of Campbell's theorem [12,35,39]. (D.M.)

Then, substituting (42.3) into (42.1) we get

$$\hat{r}(\Omega, t) = \frac{1}{[E_C d_F(t) T_{ef}]^{1/2}} \sum_{k=1}^{N} \int_{-\infty}^{\infty} Z_C(t' - t)\, y_{Fk}^*(t') \exp(j\Omega t')\, dt'. \tag{42.4}$$

Function $\hat{r}(\Omega, t)$ is a statistical quantity, because of which we must determine its average parameters, chiefly its mathematical expectation and variance. In accordance with (42.4), for the mathematical expectiation $m_r(\Omega, t) = \langle \hat{r}(\Omega, t)\rangle$ we have

$$m_r(\Omega, t) = \frac{1}{[E_C d_F(t) T_{ef}]^{1/2}} \sum_{k=1}^{N} \int_{-\infty}^{\infty} Z_C(t' - t) \langle y_{Fk}^*(t') \rangle \exp(j\Omega t')\, dt'.$$

But in view of the fact that $\langle y_{Fk}^*(t)\rangle = 0$, we find that the mathematical expectation is also equal to zero:

$$m_r(\Omega, t) = 0.$$

Next, we find the variance of the quantity (42.4)

$$d_r(\Omega, t) = \langle |\hat{r}(\Omega, t)|^2 \rangle. \tag{42.5}$$

Substituting (42.4) into (42.5) gives

$$d_r(\Omega, t) = \frac{1}{E_C d_F(t) T_{ef}} \Big\langle \sum_{k=1}^{N} \sum_{l=1}^{N} \int_{-\infty}^{\infty} \int_{-\infty}^{\infty} Z_C(t' - t) \times$$

$$\times Z_C^*(t'' - t)\, y_{Fk}^*(t')\, y_{Fl}(t'') \exp[j\Omega(t' - t'')]\, dt'\, dt'' \Big\rangle.$$

Representing the sum in the last expression as two terms (the first at $k = l$ and the second at $k \neq l$) and averaging with respect to N, we get

$$d_r(\Omega, t) = \frac{n(t)}{E_C d_F(t)} \int_{-\infty}^{\infty} \int_{-\infty}^{\infty} Z_C(t' - t)\, Z_C^*(t'' - t) \langle y_F^*(t')\, y_F(t'')\rangle \times$$

$$\times \exp[j\Omega(t' - t'')]\, dt'\, dt'' + \sum_{N=0}^{\infty} P(N, t) \frac{n(t)}{E_C d_F(t)} \times$$

$$\times \sum_{\substack{k=1 \\ k \neq l}}^{N} \sum_{l=1}^{N} \int_{-\infty}^{\infty} \int_{-\infty}^{\infty} Z_C(t' - t)\, Z_C^*(t'' - t) \langle y_{Fk}^*(t')\, y_{Fl}(t'')\rangle \exp[j\Omega(t' - t'')]\, dt'\, dt''. \tag{42.6}$$

But in view of the fact that $\langle y^*_{Fk}(t')y_{Fl}(t'')\rangle = 0$, $k \neq l$, the second term of (42.6) containing the double summation becomes zero. Therefore, we proceed from (42.6) to the following expression:

$$d_r(\Omega, t) = \frac{n(t)}{E_C d_F(t)} \int_{-\infty}^{\infty}\int_{-\infty}^{\infty} Z_C(t'-t) Z^*_C(t''-t) \times$$
$$\times \langle y^*_F(t') y_F(t'')\rangle \exp[j\Omega(t'-t'')]\, dt'dt''. \qquad (42.7)$$

Now we recall that

$$y_F(t) = \alpha\varphi(t) Z_C(t-\tau'), \qquad (42.8)$$

where τ' is the random time lag of elementary scattered signals with uniform distribution of the type

$$W_\tau(\tau') = \begin{cases} \frac{1}{T_{ef}}, & \tau' \in [t \pm T_{ef}/2], \\ 0, & \tau' \overline{\in} [t \pm T_{ef}/2]. \end{cases} \qquad (42.9)$$

Considering (42.2), (42.7)-(42.9), and assuming as previously that the function $\varphi(t)$ changes little in the interval $(t - T_{ef}/2,\ t + T_{ef}/2)$, for variance $d_r(\Omega, t)$ we get

$$d_r(\Omega, t) = \frac{1}{E^2_C T_{ef}} \int_{t-T_{ef}/2}^{t+T_{ef}/2}\int_{-\infty}^{\infty}\int_{-\infty}^{\infty} Z_C(t'-t) Z^*_C(t''-t) \times$$
$$\times Z^*_C(t'-\tau') Z_C(t''-\tau') \exp[j\Omega(t'-t'')]\, dt'\, dt''\, d\tau'. \qquad (42.10)$$

We make the following substitutions in (42.10): $\tau = \tau' - t$, $x' = t' - t$, and $x'' = t'' - t$. Ignoring edge effects we extend the integration with respect to τ' over the interval $(-\infty, \infty)$. Then, in accordance with (42.10), for $d_r(\Omega, t) = d_r(\Omega)$, we get

$$d_r(\Omega) = \frac{1}{E^2_C T_{ef}} \int_{-\infty}^{\infty}\int_{-\infty}^{\infty}\int_{-\infty}^{\infty} Z_C(x') Z_C(x'') Z^*_C(x'-\tau) \times$$
$$\times Z^*_C(x''-\tau) \exp[j\Omega(x'-x)]\, dx'\, dx''\, d\tau. \qquad (42.11)$$

Considering now that (see §22),

$$\chi(\Omega,\ \tau) = \frac{1}{E_C}\int_{-\infty}^{\infty} Z_C(t-\tau)\, Z_C^*(t)\exp(j\Omega t)\,dt \tag{42.12}$$

is the ambiguity function of the emitted signal, we get the following final expression for $d_r(\Omega)$ from (42.11):

$$d_r(\Omega) = \frac{1}{T_{ef}}\int_{-\infty}^{\infty} |\chi(\Omega,\ \tau)|^2\,d\tau. \tag{42.13}$$

As an example, let us examine a case in which the emitted signal has a bell-shaped envelope and a linear frequency modulation, for which

$$|\chi(\Omega,\ \tau)|^2 = \exp\left[-\frac{\pi}{2}\left(\frac{\tau}{T_{ef}}\right)^2 - \frac{1}{2\pi}\left(\frac{2\mu_M\tau}{T_{ef}} + \Omega T_{ef}\right)^2\right], \tag{42.14}$$

where $\mu_M = \Delta\Omega_M T_{ef}/2\pi$ is the relative deviation of the signal's frequency. Substituting (42.14) into (42.13) and designating $\gamma = \Omega T_{ef}/2\pi$, we get

$$d_r(\gamma) = \sqrt[4]{\frac{2}{1+4\mu_M^2}}\exp\left(-\frac{\pi\gamma^2}{1+4\mu_M^2}\right). \tag{42.15}$$

Let us discuss some laws which the variance $d_r(\gamma)$ obeys. It follows from (42.15) that as parameter μ_M increases -- that is, as the relative deviation of frequency in the emitted signal increases, $d_r^{1/2}(\gamma)$ decreases in a frequency range γ close to zero. At the same time we observe the following effect for $\gamma \gg 1$: As parameter μ_M increases, $d_r^{1/2}(\gamma)$ first rises, and after that it begins to decrease. This means that at any $\gamma \gg 1$ there exists such values of μ_M at which the variance $d_r(\gamma)$ is maximum.

§43. Generalization for the Case of Elementary Scattered Signals of Different Forms

In developing a discrete canonical model of reverberation (see §36) we assumed in the general case that the form of the elementary scattered signals can depend on a set of random parameters ε. This means that the probability characteristics of reverberation must be found through additional averaging with respect to ε. Thus, the instantaneous autocorrelation function of reverberation $B_F(t,\tau)$ is defined, subject to the relationship (40.8), as

$$B_F(t,\ \tau) = \langle\alpha^2\rangle\, n\,(t)\,\varphi^2\,(t) \int\limits_E \int\limits_{-\infty}^{\infty} Z_C\,(t - \tau/2,\ \varepsilon)\, Z_C^*\,(t - \tau/2,\ \varepsilon)\, W\,(\varepsilon)\, d\varepsilon\, dt, \tag{43.1}$$

where $W(\varepsilon)$ is the probability density, multidimensional in the general case, for the set of random parameters $\varepsilon = (\varepsilon_1, \varepsilon_2, \ldots, \varepsilon_M)$, and E is the domain in which these parameters exist: $\varepsilon \in E$. Correspondingly, considering (41.9), for the energy spectrum of reverberation $G_F(t,\omega)$, we have

$$G_F\,(t,\ \omega) = \langle\alpha^2\rangle\, n\,(t)\,\varphi^2\,(t) \int\limits_E |\, g_C\,(\omega,\ \varepsilon)\,|^2\, W\,(\varepsilon)\, d\varepsilon. \tag{43.2}$$

We note that we can use the parameters ε to take account of, for example, movement of scatterers, their frequency properties, and other factors defining the form of elementary scattered signals generating a reverberation process at the point of reception.

The introduction of the probability density $W(\varepsilon)$ forces us to examine the nature of the randomness of the parameters ε. In any case, we can state that these parameters reflect frequency-space-time properties of scatterers that have a statistical nature within the framework assumed for the problem.

§44. Space-Time Correlation of Reverberation when Wideband Signals Are Emitted

The case in which reverberation is observed when wideband signals are emitted has an important practical aspect. Following the classification given above (see §20), we define such signals as those of the explosive types or ones similar to them. In studying the properties of reverberation for the case under examination, we must take account of the effect of the frequency properties of the water medium and the scatterers. As in the preceding paragraphs of this chapter, we will use the discrete canonical model of reverberation.

Let $C(t)$ be a real function characterizing the form of the emitted signal. In accordance with the model of reverberation adopted here, the kth elementary signal

$f_{kr}(t, \boldsymbol{r})$, which is produced by scattering from the kth irregularitiy (with coordinates $\boldsymbol{\varrho}_k$) and is observed at the point of reception with coordinates $\boldsymbol{r}$ at time t, is defined by the relationship

$$f_x(t, \boldsymbol{r}) = \Gamma a_k \frac{C_1\left(t - \frac{|\boldsymbol{\varrho}_k| + |\boldsymbol{\varrho}_k - \boldsymbol{r}|}{c}\right)}{|\boldsymbol{\varrho}_k|\,|\boldsymbol{\varrho}_k - \boldsymbol{r}|}, \qquad (44.1)$$

where $C_1(t)$ is a function defining the form of the kth elementary scattered signal, Γ is a certain factor taking account of the energy characteristics of the emitter and receiver, a_k is a random variable defining the energetic scattering properties of the kth scatterer, and c is the rate of propagation of the acoustic wave (here we do not take account of the frequency properties of the emitter and receiver, assuming that they are known and relate to the apparatus characteristics).*

Next, we designate the spectrum of the emitted signal $C(t)$ by $g(\omega)$ and the frequency characteristics of the water medium and the kth scatterer by $K_c(\omega, \boldsymbol{\varrho}_k)$ and $K_p(\omega, \xi_k)$ (in this case ξ_k is an ensemble of random parameters on which the frequency properties of the kth scatterer may depend.† Then, obviously,

$$C_1\left(t - \frac{|\boldsymbol{\varrho}_k| + |\boldsymbol{\varrho}_k - \boldsymbol{r}|}{c}\right) = \frac{1}{2\pi} \int_{-\infty}^{\infty} g(\omega) K_c(\omega, |\boldsymbol{\varrho}_k - \boldsymbol{r}|) \times$$

$$\times K_p(\omega, \xi_k) \exp\left[j\omega\left(t - \frac{|\boldsymbol{\varrho}_k| + |\boldsymbol{\varrho}_k - \boldsymbol{r}|}{c}\right)\right] d\omega. \qquad (44.2)$$

---

*This includes the assumption of omnidirectionality of the transmitting and receiving apertures [(d), §36 above], which in turn implies that these operators be frequency-insensitive. In general, of course, these apertures are frequency-selective, particularly for broad-band inputs, see, for example [13,35,39]. [D.M.]

†Such parameters can include the resonant frequency of the scatterer, the width of its frequency characteristic, and so on.

We sum over the entire scattering domain D -- that is, we find the resultant reverberation signal

$$F(t, \mathbf{r}) = \sum_{k \in D} f_k(t, \mathbf{r}). \tag{44.3}$$

Then, substituting (44.1) and (44.2) into (44.3) we get

$$F(t, \mathbf{r}) = \frac{\Gamma}{2\pi} \sum_{k \in D} \frac{a_k}{|\boldsymbol{\varrho}_k| \, |\boldsymbol{\varrho}_k - \mathbf{r}|} \times$$

$$\times \int_{-\infty}^{\infty} g(\omega) K_c(\omega, |\boldsymbol{\varrho}_k - \mathbf{r}|) K_p(\omega, \xi_k) \exp\left[j\omega \left(t - \frac{|\boldsymbol{\varrho}_k| + |\boldsymbol{\varrho}_k - \mathbf{r}|}{c} \right) \right] d\omega. \tag{44.4}$$

Relationship (44.4) reflects the discrete canonical model of reverberation signals for this case in which wide-band signals are emitted.

Next, let us examine a particular but typical case in sonar practice when

$$\left. \begin{array}{l} |\mathbf{r}| \ll |\boldsymbol{\varrho}_k|, \\ |\boldsymbol{\varrho}_k| \gg cT_{\mathrm{ef}} \end{array} \right\} \tag{44.5}$$

The first condition in (44.5) means that the points of observation of reverberation $\mathbf{r}$ are much closer to the emitter than to the scattering domain D, while the second means that the spatial length of the emitted signal is much smaller than the distance from the emitter to domain D. We introduce the designations $\rho_k = |\boldsymbol{\varrho}_k|$, $r = |\mathbf{r}|$ and we keep in mind that under the conditions (44.5)

$$\frac{|\boldsymbol{\varrho}_k| + |\boldsymbol{\varrho}_k - \mathbf{r}|}{c} = t_k - \Delta t(r, \alpha_k, \theta_k),$$

where $t_k = 2\rho_k/c$, $\Delta t(r, \alpha_k, \theta_k)$ is the time difference in the progress of the kth elementary scattered signal between the emission point $\mathbf{r} = 0$ and point $\mathbf{r}$ at which the receiver is located, α_k and θ_k are angles in spherical coordinates describing the position of the kth scatterer, and $\boldsymbol{\varrho}_k = (\rho_k, \alpha_k, \theta_k)$. Considering this, we write (44.4) as

$$F(t,\ \boldsymbol{r})=\frac{2\,\Gamma}{\pi c^2}\sum_{k\in D}\frac{a_k}{t_k^2}\int_{-\infty}^{\infty}g(\omega)\,K_c(\omega,\ t)\,K_p(\omega,\ \xi_k)\times$$
$$\times\exp\{j\omega\,[t-t_k+\Delta t\,(r,\ \alpha_k,\ \theta_k)]\}\,d\omega. \qquad (44.6)$$

Next, we determine the space-time cross-correlation function of reverberation

$$B(t_1,\ t_2;\ \boldsymbol{r}_1,\ \boldsymbol{r}_2)\equiv\langle F(t_1,\ \boldsymbol{r}_1)\,F(t_2,\ \boldsymbol{r}_2)\rangle, \qquad (44.7)$$

in which case the averaging called for in (44.7) should be performed with respect to the random variables a_k, t_k, α_k, θ_k, and $\xi_k = \xi$, which we will consider statistically independent on the basis of the assumptions made above. Taking account of the assumptions formulated above, and of the fact that the probability characteristics of scattering irregularities do not depend on their k-number, on the basis of (44.6) and (44.7) we have

$$B(t_1, t_2; \boldsymbol{r}_1, \boldsymbol{r}_2)=\left(\frac{2\,\Gamma}{\pi c^2}\right)^2\langle a^2\rangle\int_0^{2\pi}\int_{-\pi/2}^{\pi/2}\int_{-\infty}^{\infty}\int_{-\infty}^{\infty}\int_{-\infty}^{\infty}\frac{1}{t^4}\,n(t)\times$$
$$\times\,g(\omega')\,g(\omega'')\,K_c(\omega',t)\,K_c(\omega'',t)\,\langle K_p(\omega',\xi)\,K_p(\omega'',\xi)\rangle\times$$
$$\times\exp[-j(\omega'+\omega'')\,t]\,W(\alpha,\theta)\exp[j\omega''\,\Delta t\,(r,\alpha,\theta)]\times$$
$$\times\exp[j(\omega' t_1+\omega'' t_2)]\,dt\,d\omega'\,d\omega''\cos\theta\,d\theta\,d\alpha, \qquad (44.8)$$

where $n(t)T_{ef}$ is the average density of elementary scattered signals creating reverberation during time interval $(t - T_{ef}/2,\ t + T_{ef}/2)$, $n(t)$ is the average number of elementary signals arriving at the point of reception within a unit of time, and $W(\alpha,\theta)$ is the probability density* for the location of scatterers at coordinates α_k,θ_k.

If we study the so-called "wavefront correlation", then, designating the distance between receivers located

---

*This implies that the Poisson density $\rho(\lambda)$, $c\lambda = \varrho$ here, [35,39] of scatterers factors into the product $\rho(\lambda) = \rho(\lambda)\rho_2(\alpha,\theta)$. [D.M.]

along a line passing through the origin of the coordinates (the point of emission) by r_r, we have [14]

$$\Delta t(r, \alpha, \theta) = \frac{r_r}{c}(\sin\theta\sin\theta_0 - \cos\theta\cos\theta_0\sin\alpha).$$

Let us examine some particular cases of the correlation properties of reverberation signals. We set the frequency properties of the medium to be

$$K_c(\omega, t) = \exp[-0.1c\beta(\omega)t], \tag{44.9}$$

where $\beta(\omega)$ is the absorption coefficient (in nepers per unit distance). Such a representation of $K_c(\omega,t)$ corresponds to a marine environment with constant absorption per unit distance in the direction of signal propagation. We assume that the nonstationariness of reverberation is such that functions $1/t^4$, $n(t)$, and $K_c(\omega,t)$ change insignificantly over the interval $(t - T_{ef}/2,\ t + T_{ef}/2)$. Then, integrating with respect to t in (44.8) we get

$$\int_{-\infty}^{\infty} \frac{1}{t^4} n(t) K_c(\omega', t) K_c^*(\omega'', t) \exp[-j(\omega' + \omega'')t]\, dt =$$

$$= \frac{2\pi}{t^4} n(t) K_c(\omega', t) K_c^*(\omega'', t)\, \delta(\omega' + \omega''), \tag{44.10}$$

where $\delta(\omega)$ is a delta-function. Introducing the transformations $\omega = \omega' - \omega''$, $\tau = t_2 - t_1$, and considering (44.9) and (44.10), we get the following for the correlation function $B(t_1,t_2;\ \mathbf{r}_1,\mathbf{r}_2) = B(t,\tau;\mathbf{r}_r)$ as defined in (44.8):

$$B(t, \tau; r_r) = \frac{8\Gamma^2 \langle a^2 \rangle n(t)}{\pi c^4 \quad t^4} \int_0^{2\pi} \int_{-\pi/2}^{\pi/2} \int_0^{\infty} |g(\omega)|^2 \langle |K_p(\omega, \xi)|^2 \rangle \times$$

$$\times \exp[-0.2c\beta(\omega)t] \cos\{\omega[\tau + \Delta t(r_r, \alpha, \theta)]\}\, W(\alpha, \theta)\, d\omega \cos\theta\, d\theta\, d\alpha. \tag{44.11}$$

As follows from (44.11), despite assumption (44.10) the reverberation process is essentially nonstationary, since not only the variance but also the form of the correlation function depends on the current time t. We note that when wideband signals are emitted, reverberation is a process which can be reduced to a stationary state only in the event that sound absorption in the working frequency band changes insignificantly.

Using expression (44.11) we define the temporal $B(t,\tau;\ 0)$ and spatial $B(t,0;\ r_r)$ correlation functions of reverberation:

$$B(t,\ \tau;\ 0) \equiv \frac{8\,\Gamma^2}{\pi c^4}\frac{\langle a^2\rangle\, n(t)}{t^4}\int\limits_0^\infty |g(\omega)|^2 \langle |K_p(\omega,\ \xi)|^2\rangle \exp[-0.2c\beta(\omega)\,t] \cos\omega\tau\, d\omega; \tag{44.12}$$

$$B(t,\ 0;\ r_r) \equiv \frac{8\Gamma^2}{\pi c^4}\frac{\langle a^2\rangle\, n(t)}{t^4}\int\limits_0^{2\pi}\int\limits_{-\pi/2}^{\pi/2}\int\limits_0^\infty |g(\omega)|^2 \langle |K_p(\omega,\ \xi)|^2\rangle \times$$

$$\times \exp[-0.2c\beta(\omega)\,t] \cos[\omega\,\Delta t\,(r_r,\ \alpha,\ \theta)]\, W(\alpha,\ \theta)\, d\omega \cos\theta\, d\theta\, d\alpha. \tag{44.13}$$

It follows from (44.12) and (44.13), in particular, that the spectrum of the emitted signals as well as the frequency characteristics of scatterers and the marine environment affect both the temporal and the spatial correlation of reverberation.

Let us now examine some specific distributions of scatterers in space typical of reverberation signals observed under various sonar conditions.

If scattering occurs at irregularities within a boundless medium and if the average density of the scatterers is constant, then

$$W(\alpha,\ \theta) = \begin{cases} W(\alpha)\,W(\theta) = \dfrac{1}{2\pi}\,\dfrac{1}{\pi}, & \alpha \in [0,2\pi], \quad \theta \in \left[-\dfrac{\pi}{2},\ \dfrac{\pi}{2}\right]; \\ 0, & \alpha \bar{\in} [0,2\pi], \quad \theta \bar{\in} \left[-\dfrac{\pi}{2},\ \dfrac{\pi}{2}\right]. \end{cases} \tag{44.14}$$

This case pertains to observation of three-dimensional (in volume) reverberation caused by the scattering of sound by independent, discrete irregularities. For simplicity we assume in (44.10) that $\theta_0 = \pi/2$, such that $\Delta t(r_r,\alpha,\theta) = \Delta t(r_r,\theta) = (r_r/c)\sin\theta$. Then from (44.11) and (44.14) we get

$$B(t, \tau; r_r) = \frac{4\Gamma^2}{\pi^3 c^4} \frac{\langle a^2 \rangle n(t)}{t^4} \int_0^\infty |g(\omega)|^2 \langle | K_p(\omega, \xi)|^2 \rangle \times$$

$$\times \exp[-0.2c\beta(\omega) t] \frac{\sin \frac{\omega r_r}{c}}{\frac{\omega r_r}{c}} \cos \omega\tau \, d\omega. \quad (44.15)$$

This case corresponds to the use of a nondirectional (or "omnidirectional") emitter.*

It follows from (44.15) that the spatial correlation of reverberation does not depend on the orientation of receivers in space and is governed only by the distance r_r between them. Correlation of this type is typical of so-called diffuse random fields.

If reverberation is caused by scattering at irregularities distributed as a thin layer (or on a boundary separating two media) and the emitter and receiver are located near the scatterers, then

$$W(\alpha, \theta) = W(\alpha) W(\theta) = \frac{1}{2\pi} \delta(\theta_0), \ \alpha \in [0, 2\pi]. \quad (44.16)$$

After (44.16) is substituted into relationship (44.11) the latter takes the form:

$$B(t, \tau; r_r) = \frac{8\Gamma^2}{\pi c^4} \frac{\langle a^2 \rangle n(t)}{t^4} \int_0^\infty |g(\omega)|^2 \langle | K_p(\omega, \xi)|^2 \rangle \times$$

$$\times \exp[-0.2c\beta(\omega) t] J_0\left(\frac{\omega r_r}{c} \cos\theta_0\right) \cos \omega\tau \, d\omega, \quad (44.17)$$

where $J_0(x)$ is a zero-order Bessel function of real argument. This case pertains to the observation of bottom,

---

*As postulated, cf. §36; also, remarks following Eq. (44.1). [D.M.]

and in some cases surface, reverberation. It follows from (44.17) in particular, that when the angle θ_0 is changed from zero to $\pi/2$ the correlation function of reverberation increases. This means that correlation "along the wave-front" in the plane of the scatterers is lower than in directions at an angle to this plane. It should be kept in mind that relationship (44.17) corresponds to an omnidirectional emitter.

§45. One-Dimensional Semi-invariants

Let us examine the model of reverberation signals

$$F(t) = \sum_{k=1}^{N} \alpha_k \varphi(t_k) C(t - t_k) \tag{45.1}$$

and define one-dimensional semi-invariants $\lambda_m(t)$, of the mth order. These are factors in the expansion of the logarithm of the characteristic function

$$\ln \Theta(\eta, t) = \sum_{m=1}^{\infty} \frac{(j\eta)^m}{m!} \lambda_m(t). \tag{45.2}$$

A knowledge of the semi-invariants permit us to determine the probability distribution laws for reverberation by approximation and, in particular, to evaluate the effect of the limiting number of elementary scattered signals and of coherent scattering on these distributions (see §46).

Considering the independence of elementary scattered signals, we have

$$\Theta(\eta, t \mid N) = \Theta^N(\eta, t \mid 1), \tag{45.3}$$

where $\Theta(\eta, t|N)$ and $\Theta(\eta, t|1)$ are conditional characteristic functions for the sum N and one of the scattered signals. By definition,

$$\Theta(\eta, t \mid 1) = \langle \exp[j\alpha_k \varphi(t_k) C(t - t_k)] \rangle \tag{45.4}$$

and, moreover,

$$\Theta(\eta, t) = \sum_{N=1}^{\infty} P(N, t)\, \Theta^N(\eta, t \mid 1), \tag{45.5}$$

where $P(N,t)$ is the probability distribution for N scattered signals.

We take account of the assumptions made in §38 and of the fact that here

$$P(N,t)=\frac{[n(t)T]^N}{N!}\exp[-n(t)T] \tag{45.6}$$

corresponds to a Poisson distribution, where $n(t)$ is the average number of elementary scattered signals created in a unit of time, and T is the interval under consideration, in which exactly N signals arise.

At low values of the parameter $n(t)T$ the Poisson distribution has limited application.* As has been demonstrated in [14], it can be used when the condition

$$n(t)\,T>3;$$

is satisified. Otherwise, the more general Bernoulli's law is valid for $P(N,t)$.

Use of expressions (45.1), (45.3)-(45.6) leads to the following representation of the logarithm of the characteristic function of reverberation [14]:

$$\ln\Theta(\eta,t)=\sum_{m=1}^{\infty}\frac{(j\eta)^m}{m!}\langle\alpha^m\rangle n(t)\varphi(t)\int_{-\infty}^{\infty}C^m(t')\,dt'. \tag{45.7}$$

Now, comparing (45.2) and (45.7) we get the following expression for semi-invariants of the mth order:

$$\lambda_m(t)=\langle\alpha^m\rangle n(t)\int_{-\infty}^{\infty}C^m(t')\,dt'. \tag{45.8}$$

As distribution parameters, semi-invariants (45.8) have a number of properties that make their use convenient, and they can be determined simply enough on the basis of the characteristics of the emitted signals and scatterers. Let us dwell on some properties of semi-invariants which we will use subsequently.

---

*However, see [52]. (D.M.)

One of the properties of semi-invariants is that they are additive. If the random variable $y=\sum_{i=1}^{N} x_i$, equal to the sum of independent random variables x_i, is given, then the semi-invariant $\lambda_k^{(y)}$ of any order k for variable y is equal to the sum of the semi-invariants $\lambda_k^{(x_i)}$ of the terms -- that is,

$$\lambda_k^{(y)}=\sum_{i=1}^{N}\lambda_k^{(x_i)}. \tag{45.9}$$

Semi-invariants uniquely define the initial moments $\alpha_k = \langle x^k\rangle$ and central moments $\mu_k = \langle (x-\langle x\rangle)^k\rangle$ of the distribution. In particular, the following relationships are valid for these distribution parameters:

$$\left.\begin{aligned} &\lambda_1=\alpha_1;\\ &\lambda_2=\alpha_2-\alpha_1^2;\\ &\lambda_3=\alpha^3-3\alpha_1\alpha_2+2\alpha_1^3;\\ &\lambda_4=\alpha_4-3\alpha_2^2-4\alpha_1\alpha_3+12\alpha_1^2\alpha_2-6\alpha_1^4;\\ &\ldots\ldots\ldots\ldots\ldots\ldots \end{aligned}\right\} \tag{45.10}$$

$$\left.\begin{aligned} &\alpha_1=\lambda_1;\\ &\alpha_2=\lambda_2-\lambda_1^2;\\ &\alpha_3=\lambda_3+3\lambda_1\lambda_2+\lambda_1^3;\\ &\alpha_4=\lambda_4+3\lambda_2^2+4\lambda_1\lambda_3+6\lambda_1^2\lambda_2+\lambda_1^4;\\ &\ldots\ldots\ldots\ldots\ldots\ldots \end{aligned}\right\} \tag{45.11}$$

$$\left.\begin{aligned} &\lambda_1=\alpha_1;\\ &\lambda_2=\mu_2;\\ &\lambda_3=\mu_3;\\ &\lambda_4=\mu_4-3\mu_2^2;\\ &\ldots\ldots \end{aligned}\right\} \tag{45.12}$$

It follows, in particular from (45.10)-(45.12), that the values of such distribution parameters as the coefficients of variation γ_v, asymmetry, γ_a, and excess, γ_e, are expressed in the following way by the semi-invariants λ_k, initial moments α_k, and central moments μ_k:

$$\gamma_v = \frac{\lambda_2^{1/2}}{\lambda_1} = \frac{(\alpha_2 - \alpha_1)^{1/2}}{\alpha_1} = \frac{\mu_2^{1/2}}{\alpha_1}; \qquad (45.13)$$

$$\gamma_a = \frac{\lambda_3}{\lambda_2^{3/2}} = \frac{\alpha_3 - 3\alpha_1\alpha_2 + 2\alpha_1^3}{(\alpha_2 - \alpha_1^2)^{3/2}} = \frac{\mu_3}{\mu_2^{3/2}}; \qquad (45.14)$$

$$\gamma_e = \frac{\lambda_4}{\lambda_2^2} = \frac{\alpha_4 - 3\alpha_2^2 - 4\alpha_1\alpha_3 + 12\alpha_1^2\alpha_3 - 6\alpha_1^4}{(\alpha_2 - \alpha_1^2)^2} = \frac{\mu_4}{\mu_2^2} - 3. \qquad (45.15)$$

§46. Probability Distributions for Instantaneous Values of Reverberation and Its Envelope

Expression (45.8) obtained above permits us to analyze the probability distributions of reverberation signals. We will examine this problem under the following two assumptions:

1. the average number of elementary scattered signals $n(t)$ is arbitrary and, in particular, rather small;

2. there is a quasideterministic component (coherent scattering) in the reverberation process.

Thus, the reverberation signal $F(t)$ is represented by

$$F(t) = F_1(t) + F_2(t), \qquad (46.1)$$

where $F_1(t)$ is the random component of reverberation and $F_2(t)$ is the quasideterministic component.

We will subsequently need a representation for reverberation signals based on the possibility of defining random functions as the product of two factors, namely envelope $E(t)$ and the cosine of the instantaneous phase, $\omega_0 t + \Psi(t)$, where ω_0 is the central frequency of the process's spectrum. If we ignore regular movement of scatterers in the marine environment, and differences in sound absorption within the frequency band of the emitted signals, then the following representation is valid for $F(t)$:

$$F(t) = E(t) \cos [\omega_0 t + \Psi(t)] + E_0(t) \cos (\omega_0 t + \Psi_0), \quad (46.2)$$

where $E_0(t)$ and Ψ_0 are the envelope and the initial phase of the quasideterministic component of the reverberation. It is not difficult to see, on assuming for the sake of simplicity that $E_0(t) = E_0 = \text{const}$ and $\Psi_0 = 0$ -- that is, considering the amplitude of $E_0(t)$ to be constant for a certain time t and the initial phase to be equal to zero -- that we then get the following equivalent relationship from (46.2):

$$F(t) = [E(t) \cos \Psi(t) + E_0] \cos \omega_0 t - E(t) \sin \Psi(t) \sin \omega_0 t, \quad (46.3)$$

where

$$\left.\begin{aligned} E(t) \cos \Psi(t) + E_0 &= F_c(t); \\ E(t) \sin \Psi(t) &= F_s(t) \end{aligned}\right\} \quad (46.4)$$

are the composite components of reverberation.* For narrow-band emitted signals $C(t)$, random functions $E(t)$ and $\Psi(t)$ change slowly with time compared to cos $\omega_0 t$ and sin $\omega_0 t$, making it possible to isolate them using simple technical resources [9.14].

It should be kept in mind that representations (46.2)-(46.4) are valid for both narrow-band and wideband signals. In the latter case we can obtain the envelope and phase by introducing the conjugate process $\tilde{F}(t)$, which is a Hilbert transform of the initial process $F(t)$:

$$\tilde{F}(t) = \frac{1}{\pi} \int_{-\infty}^{\infty} \frac{F(t')\, dt'}{t - t'}. \quad (46.5)$$

As we know, integration components $F_c(t)$ and $F_s(t)$ are associated with each other by a transformation of the type (46.5) and can be obtained directly from $F(t)$ and $\tilde{F}(t)$. In this case it is necessary to perform a frequency trans-

---

*Functions $F_c(t)$ and $F_s(t)$ should be called the envelopes of the composite components, since they define the complex envelope of reverberation $Z_F(t) = F_c(t) + j\tilde{F}_s(t)$.

formation, such that the central frequency ω_0 falls at point $\omega_0 = 0$.

Let us go on to an examination of the probability densities $W(F,t)$ and $W(F_c,F_s,t)$ of the reverberation and its envelope components $F_c(t)$ and $F_s(t)$, taking account of low scatterer density in the marine environment and the presence of a quasideterministic component (coherent scattering).

First, we find the distribution $W(F_1,t)$ of the Poisson component $F_1(t)$ of reverberation.* This distribution can be found approximately with the known semi-invariants (45.8) of the reverberation process by representing $W(F_1,t)$ as an Edgeworth series. An Edgeworth series permits us to define the probability density of a certain random variable if we know the distribution moments or the semi-invariants. One of the ways to write an Edgeworth series is:

$$W(F_1,t)\,\sigma_F(t) = \frac{1}{\sqrt{2\pi}}\exp\left(-\frac{x^2}{2}\right)\left[1 - \frac{1}{3!}\frac{\mu_3(t)}{\sigma_F^3(t)}H_3(x) + \right.$$
$$+\frac{1}{4!}\left(\frac{\mu_4(t)}{\sigma_F^4(t)} - 3\right)H_4(x) + \frac{10}{6!}\left(\frac{\mu_3(t)}{\sigma_F^3(t)}\right)^2 H_6(x) -$$
$$\left. -\frac{1}{5!}\left(\frac{\mu_5(t)}{\sigma_F^5(t)} - 10\frac{\mu_3(t)}{\sigma_F^3(t)}\right)H_5(x) + \cdots\right], \qquad x = \frac{F_1}{\sigma_F(t)}, \tag{46.6}$$

where $\mu_k(t)$ are the central moments of the (here) Poisson component of reverberation, $H_m(x)$ are mth order Hermite polynomials, and $\sigma_F^2(t) = \mu_2(t)$.

If we are using narrow-band emitted signals $C(t)$ -- that is, if

$$C(t) = C_0(t)\cos\omega_0 t, \tag{46.7}$$

then on substituting (46.7) into (45.8) it follows that

$$\lambda_m(t) = 0, \qquad m = 1, 3, 5, \ldots \tag{46.8}$$

---

*[35]; Chapter 7, [12]. (D.M.)

This means that all odd-order central moments are also equal to zero:

$$\mu_m(t) = 0, \qquad m = 1, 3, 5, \ldots \tag{46.9}$$

Then, considering relationships (45.8), (45.10)-(45.15), and (46.6)-(46.9), we get the following approximate relationship for $W(F,t)$:

$$W(F_1, t)\,\sigma_F(t) \approx \frac{1}{2\pi}\exp\left(-\frac{x^2}{2}\right)\left[1 + \frac{\gamma_э(t)}{4!}(x^4 - 6x^2 + 3)\right], \tag{46.10}$$

where [14]

$$\gamma_э(t) = \frac{\langle\alpha^4\rangle}{\langle\alpha^2\rangle^2 n(t)}\,\frac{\int\limits_{-\infty}^{\infty} C^4(t')\,dt'}{\left[\int\limits_{-\infty}^{\infty} C^2(t')\,dt'\right]^2} \tag{46.11}$$

is the coefficient of excess of the reverberation, depending, as can be seen from (46.11), on the distribution $W(\alpha)$, the forms of the emitted signals $C(t)$, and on the average number $n(t)$ of scattered signals arriving at the point of reception within a unit of time -- that is, on the average density of scatterers in the marine environment. It is easy to see that as $n(t) \to \infty$ -- that is, as $\gamma \to 0$, probability density $W(F_1,t)$ converges to a normal distribution

$$W(F_1, t)\,\sigma_F(t) = \frac{1}{\sqrt{2\pi}}\exp\left(-\frac{x^2}{2}\right), \tag{46.12}$$

which agrees with the central limit theorem of probability theory, according to which the sum of a large number of independent terms takes on a normal distribution as a limit, independent of the type of distribution the terms have.*

On the basis of expression (46.2), the quasideterministic component $F_2(t)$ of reverberation can be defined as a harmonic process with a random initial phase. Then, its density distribution $W(F_2)$ is defined by the known relationship

---

*[12], Chapter 7. (D.M.)

$$W(F_2)=\frac{1}{\pi E_0(1-Y^2)^{1/2}},\quad |Y|\leqslant 1,\ Y=\frac{F_2}{E_0}, \tag{46.13}$$

where the phase Ψ is uniformly distributed over the interval $(-\pi,\pi)$, and where E_0 is the amplitude of the quasideterministic component of the reverberation. Now we can find the distribution $W(F,t)$ of the total reverberation signal $F(t)$, as defined by (46.1). It is obvious that both components of the reverberation, $F_1(t)$ and $F_2(t)$, are statistically independent, since the random and quasideterministic components are generated by the scattering of sound off of different irregularities in the medium. Then the following relationship is valid for $W(F)$:

$$W(F,\ t)=\int_{-\infty}^{\infty} W(F_1=F-F',\ t)\,W(F_2=F')\,dF', \tag{46.14}$$

where integrands $W(F_1)$ and $W(F_2)$ were defined by formulas (46.10) and (46.13), respectively, after corresponding substitution of the arguments. Taking this into account, we get the following for $W(F,t)$:

$$W(F,\ t)\,\sigma_F(t)\approx\frac{1}{\pi\sqrt{2\pi}\,E_0}\int_{-1}^{1}\frac{\exp\left[-\frac{(F,t-F,t')^2}{2\sigma_F^2(t)}\right]}{\left[1-\left(\frac{F'}{E_0}\right)^2\right]^{1/2}}\times$$

$$\times\left\{1+\frac{\gamma_e(t)}{4!}\left[\left(\frac{F-F'}{\sigma_F(t)}\right)^4-6\left(\frac{F-F'}{\sigma_F(t)}\right)^2+3\right]\right\}dF'. \tag{46.15}$$

Then, after we make the substitution $F' = E\cos\Theta$ for the variable of integration in expression (46.15), we get

$$W(F,\ t)\,\sigma_F(t)\approx\frac{1}{\pi\sqrt{2\pi}}\int_0^{\pi}\exp\left[-\frac{(z-q(t)\cos\theta)^2}{2}\right]\times$$

$$\times\left\{1+\frac{\gamma_e(t)}{4!}\left[(z-q(t)\cos\theta)^4-6(z-q(t)\cos\theta)^2+3\right]\right\}d\theta, \tag{46.16}$$

$$z=\frac{F}{\sigma_F(t)},\quad q(t)=\frac{E_0}{\sigma_F(t)},$$

where $q(t)$ is a parameter characterizing the relationship between the levels of the quasideterministic component of reverberation and its random component. As we would expect, at $q(t) = 0$ the distribution $W(F,t)$ converges to a distribution of the (46.10) type, and if $\gamma_e(t) \to 0$ at the same time, then we get a normal distribution (46.12) as the limit. As $\gamma_e(t) \to 0$ and with $q(t) > 0$, formula (46.16) becomes the well-known expression [12]

$$W(F,\ t)\,\sigma_F(t) = \frac{1}{\pi\sqrt{2\pi}} \int_0^{\pi} \exp\left[-\frac{(z - q(t)\cos\theta)^2}{2}\right] d\theta \qquad (46.17)$$

for the probability density of the sum of a narrow-band normal random process and a sinusoidal signal with a random initial phase. Integrating this expression produces $W(F,t)\sigma_F(t)$ in different series form. One such expression has the form

$$W(F,\ t)\,\sigma_F(t) = \frac{1}{\sqrt[3]{2\pi}} \sum_{k=0}^{\infty} \frac{(-1)^n}{n!} \left(\frac{z^2}{2}\right)^n {}_1F_1\left(n + \frac{1}{2},\ 1,\ -\frac{q^2(t)}{2}\right), \qquad (46.18)$$

where ${}_1F_1(\alpha,\beta,\gamma)$ is a confluent hypergeometric function.

Using relationships (46.10) and (46.18) we can evaluate the effect of parameters $\gamma_e(t)$ and $q(t)$ on a number of reverberation signal distributions. It follows from the relationships derived above that a small number of scattered signals, which according to (46.11) cause $\gamma_e(t)$ to increase, "constrict" $W(F,t)$ as compared to a normal distribution. On the other hand an increase in the level of the quasideterminate component in the reverberation process causes the distribution $W(F,t)$ to "widen" as compared to a normal distribution -- that is, its coefficient of excess increases.

Given arbitrary values for the parameters $\gamma_e(t)$ and $q(t)$, analytical computations of the distribution $W(F,t)$ using formula (46.16) are difficult in the general case, although this expression can be presented in series form. For this purpose the exponent in the integrand of (46.16) may be expanded in a power series:

$$\exp\left[-\frac{(z - q(t)\cos\theta)^2}{2}\right] = \sum_{n=0}^{\infty} \frac{(-1)^n}{2^n n!} (z - q(t)\cos\theta)^{2n}. \qquad (46.19)$$

Next, using (46.16) and (46.19), for $W(F,t)$ we get

$$W(F,\ t)\,\sigma_F(t) \approx \frac{1}{\pi\sqrt{2\pi}} \sum_{n=0}^{\infty} \frac{(-1)^n}{2^n n!} \left[\left(1 + \frac{\gamma_e(t)}{8}\right) \int_0^{\pi} (z - q(t)\cos\theta)^{2n}\, d\theta - \right.$$

$$\left. - \frac{\gamma_e(t)}{4} \int_0^{\pi} (z - q(t)\cos\theta)^{2n+2}\, d\theta + \frac{\gamma_e(t)}{24} \int_0^{\pi} (z - q(t)\cos\theta)^{2n+4}\, d\theta \right]. \quad (46.20)$$

Expression (46.20) contains integrals of the form

$$J(z,\ q(t)) = \int_0^{\pi} (z - q(t)\cos\theta)^{2n}\, d\theta. \quad (46.21)$$

Representing the integrand function in (46.21) as a binomial series, we have

$$J(z,\ q(t)) = z^{2N} \sum_{k=0}^{2N} (-1)^k \binom{2N}{k} \left(\frac{q(t)}{z}\right)^k \int_0^{\pi} \cos^k\theta\, d\theta,$$

where $\binom{2N}{k}$ is the number of combinations of $2N$ with respect to k. Considering further that

$$\int_0^{\pi} \cos^k\theta\, d\theta = \begin{cases} \pi \dfrac{(k-1)!!}{k!!}, & k = 0,\ 2,\ 4,\ \ldots, \\ 0\quad , & k = 1,\ 3,\ 5,\ \ldots, \end{cases} \quad (46.22)$$

we can write the following expression for $J(z,q(t))$:

$$J(z,\ q(t)) = \pi z^{2N} \sum_{k=0}^{2N} \binom{2N}{k} \frac{(k-1)!!}{k!!} \left(\frac{q(t)}{z}\right)^k. \quad (46.23)$$

Thus, the distribution $W(F,t)$ is represented as the following binary series on the basis of relationships (46.20)-(46.23):

$$W(F,\ t)\,\sigma_F(t) = \frac{1}{\sqrt{2\pi}} \sum_{n=0}^{\infty} \frac{(-1)^n}{2^n n!} z^{2n} \left(1 + \frac{\gamma_e(t)}{8}\right) \times$$

$$\times \left[\sum_{k=0}^{2n} \binom{2n}{k} \frac{(k-1)!!}{k!!} \left(\frac{q(t)}{z}\right)^k - \frac{\gamma_e(t)}{4} z^2 \sum_{k=0}^{2n+2} \binom{2n+2}{k} \times \right.$$

$$\left. \times \frac{(k-1)!!}{k!!} \left(\frac{q(t)}{z}\right)^k + \frac{\gamma_e(t)}{24} z^4 \sum_{k=0}^{2n+4} \binom{2n+4}{k} \frac{(k-1)!!}{k!!} \left(\frac{q(t)}{z}\right)^k \right].$$

(46.24)

Further computation of $W(F,t)$ can be performed by summing the series in (46.24) with respect to k and n, for different values of the parameters $\gamma_e(t)$ and $q(t)$.

The relationships obtained above permit us to analyze the probability distributions of the reverberation envelope.

We write an expression for the two-dimensional probability density $W(F_c,F_s,t)$ of the envelope components of reverberation. We assume that the envelope components $F_c(t)$ and $F_s(t)$ are statistically independent, and that a distribution law such as (46.10) is valid for each of them when a quasideterministic component is absent from the reverberation process -- that is,

$$W(F_c,\ t)\,\sigma_F(t) = \frac{1}{\sqrt{2\pi}}\exp\left(-\frac{x_c^2}{2}\right)\left[1+\frac{\gamma_e(t)}{4!}(x_c^4-6x_c^2+3)\right], \tag{46.25}$$

$$x_c = \frac{F_c}{\sigma_F(t)},$$

$$W(F_s,\ t)\,\sigma_F(t) = \frac{1}{\sqrt{2\pi}}\exp\left(-\frac{x_s^2}{2}\right)\left[1+\frac{\gamma_e(t)}{4!}(x_s^4-6x_s^2+3)\right], \tag{46.26}$$

$$x_s = \frac{F_s}{\sigma_F(t)}.$$

Taking account of (46.25) and (46.26), and of the fact that in accordance with the assumptions made, e.g., $W(F_c,F_s,t) = W(F_c,t)W(F_s,t)$, we get

$$W(F_c,\ F_s,\ t)\,\sigma_F^2(t) \approx \frac{1}{\sqrt{2\pi}}\exp\left(-\frac{x_c^2+x_s^2}{2}\right)\times$$
$$\times\left[1+\frac{\gamma_e(t)}{4!}(x_c^4+x_s^4-6x_c^2-6x_s^2+6)\right], \tag{46.27}$$

in which case the right-hand side of (46.27) includes terms containing the multiplier $\gamma_e(t)/4!$, while terms containing the multiplier $(\gamma_e(t)/4!)^2$ have been dropped, such that this relationship be sufficiently accurate for $\gamma_e(t) < 5$.

If $\gamma_e(t) \to 0$, then we get the following two-dimensional normal probability density (p.d.) as a limit for $W(F_c,F_s,t)$:

$$W(F_c,\ F_s,\ t)\,\sigma_F^2(t) = \frac{1}{2\pi}\exp\left(-\frac{x_c^2+x_s^2}{2}\right), \tag{46.28}$$

which corresponds to a two-dimensional Gaussian distribution and to an extremely large number of scattered signals generating a reverberation process, that is, to the case in which $n(t) \to \infty$.

We note that the expression (46.20) for the two-dimensional probability density was derived on the condition that the envelopes of the integration components $F_c(t)$ and $F_s(t)$ are statistically independent. In terms of Gaussian distributions or ones similar to them for the integration components, this independence is understood in the sense of their linear independence -- that is, absence of correlation between them. On the other hand, $F_c(t)$ and $F_s(t)$ are related to each other by a Hilbert transform

$$F_c(t) = \frac{1}{\pi} \int_{-\infty}^{\infty} \frac{F_s(t')\,dt'}{t-t'},$$

that is, functionally. This should not be suprising. Let us find the cross-correlation function:

$$B_{cs} = \langle F_c(t) F_s(t) \rangle = \frac{1}{\pi} \int_{-\infty}^{\infty} \frac{\langle F_s(t) F_c(t') \rangle\,dt'}{t-t'}.$$

Since from our standpoint reverberation is a random process here reduced to a stationary state, we can write

$$\langle F_s(t) F_s(t') \rangle = d_s(t) R_s(t-t'),$$

where $d_s(t)$ is the variance and $R_s(t - t')$ is the (normalized) autocorrelation function (an even function). Then

$$B_{cs} = -\frac{d_s(t)}{\pi} \int_{-\infty}^{\infty} \frac{R_s(\tau)\,d\tau}{\tau} = 0.$$

When a quasideterministic component is present, the probability density $W(F_c, F_s, t)$ is equivalent to (46.27), if we replace x_c by $x_c - q(t)$, $q(t) = E_0/\sigma_F(t)$ in that relationship. This follows from expressions (46.3) and (46.4), in which the quasideterministic component shifts the process $F_c(t)$ by the amount E_0. Thus, we have

$$W(F_c, F_s, t)\,\sigma_F^2(t) \approx \frac{1}{2\pi} \exp\left[-\frac{(x_c - q(t))^2 + x_s^2}{2}\right] \times$$
$$\times \left\{1 + \frac{\gamma_э(t)}{4!}\left[(x_c - q(t))^4 + x_s^4 - 6(x_c - q(t))^2 - 6x_s + 6\right]\right\}. \quad (46.29)$$

The relationship (46.29) enables us to find the probability density $W(E,t)$ of the reverberation envelope $E(t)$ when the density of scatterers in the marine environment is low and when a quasideterministic component is present in the reverberation process. For this purpose we first compute the two-dimensional joint probability density $W(E,\Psi,t)$ of the envelope $E(t)$ and instantaneous phase $\Psi(t)$. As is known, the following general relationship is valid for $W(E,\Psi,t)$:

$$W(E, \Psi, t) = EW(F_c = E\cos\Psi,\ F_s = E\sin\Psi,\ t), \quad E \geqslant 0,\ |\Psi| \leqslant \pi, \tag{46.30}$$

where the righthand side represents the probability density $W(F_c,F_s,t)$ given the substitution of variables indicated in parentheses. Using (46.29) and (46.30), we get

$$W(E, \Psi, t)\,\sigma_F(t) \approx \frac{U}{2\pi}\exp\left[-\frac{(U\cos\Psi - q(t))^2 + U^2\sin^2\Psi}{2}\right]\times$$
$$\times\left\{1 + \frac{\gamma_e(t)}{4!}\left[(U\cos\Psi - q(t))^4 + U^4\sin^4\Psi -\right.\right.$$
$$\left.\left. - 6(U\cos\Psi - q(t))^2 - 6U^2\sin^2\Psi + 6\right]\right\}, \quad U = \frac{E}{\sigma_F(t)}.$$

Next, we express the probability density $W(E,\Psi,t)$ as:

$$W(E, \Psi, t)\,\sigma_F(t) \approx \frac{U}{2\pi}\exp\left(-\frac{U^2 + q^2(t) - 2Uq(t)\cos\Psi}{2}\right)\times$$
$$\times\left\{1 + \frac{\gamma_e(t)}{4!}\left[U^4\cos^4\Psi - 4U^3q(t)\cos^3\Psi + 6U^2q^2(t)\cos^2\Psi -\right.\right.$$
$$\left.\left. - 4Uq^3(t)\cos\Psi + U^4\sin^4\Psi + 12Uq(t)\cos\Psi + q^4(t) - 6q^2(t) - 6U^2 + 6\right]\right\}. \tag{46.31}$$

Performing the corresponding trigonometric transformation in expression (46.31), we find that

$$W(E, \Psi\ t)\,\sigma_F(t) \approx \frac{U}{2\pi}\exp\left(-\frac{U^2 + q^2(t) - 2Uq(t)\cos\Psi}{2}\right)\times$$
$$\times\left\{1 + \frac{\gamma_e(t)}{4!}\left[\frac{U^4}{4}\cos 4\Psi + \frac{3}{4}U^4q^4(t) - U^3q(t)\cos 3\Psi +\right.\right.$$
$$+ (12Uq(t) - 4Uq^3(t) - 3U^3q(t))\cos\Psi + 3U^2q^2(t)\cos 2\Psi +$$
$$\left.\left. + 3U^2q^2(t) - 6U^2 - 6q^2(t) + 6\right]\right\}. \tag{46.32}$$

The probability density $W(E,t)$ is derived from $W(E,\Psi,t)$ by integrating the latter with respect to Ψ within the range from $-\pi$ to π, that is,

$$W(E, t) = \int_{-\pi}^{\pi} W(E, \Psi, t)\, d\Psi. \tag{46.33}$$

Substituting expression (46.32) into (46.33) and considering that

$$I_n(x) = \frac{1}{\pi}\int_{0}^{\pi} \exp(x\cos\theta)\cos(n\theta)\, d\theta$$

is a modified nth order Bessel function, we get

$$W(E, t)\,\sigma_F(t) \approx U\exp\left(-\frac{U^2+q^2(t)}{2}\right)\Big\{\Big[1+\frac{\gamma_e(t)}{4!}\Big(\frac{3}{4}U^4+q^4(t)+ 3U^2q^2(t)-6U^2-6q^2(t)+6\Big)\Big]\, I_0(Uq(t)) + \frac{\gamma_e(t)}{4!}\,[12Uq(t)-4Uq^3(t)-3U^3q(t)\, I_1(Uq(t))] + \frac{\gamma_e(t)}{8}\,U^2q^2(t)\, I_2(Uq(t)) - \frac{\gamma_e(t)}{4!}\,U^3 I_3(Uq(t)) + \frac{\gamma_e(t)}{4\cdot 4!}\,U^4 I_4(Uq(t))\Big\}. \tag{46.34}$$

Let us examine particular cases representing the probability density (46.34) for the envelope of reverberation. If $\gamma_e(t) \to 0$ -- that is, if the number of scatterers in the medium is extremely large, then from (46.34) we get the so-called generalized Rayleigh law* as a limit:

$$W(E, t)\,\sigma_F(t) = U\exp\left(-\frac{U^2+q^2(t)}{2}\right) I_0(Uq(t)). \tag{46.35}$$

An analysis of the relationship (46.35) indicates, in particular, that as the parameter $q(t)$ increases, the probability density $W(E,t)$ shifts to the right and converges at $q(t) \gg 1$ to a normal distribution. This means that the coefficients of variation, $\gamma_v^{(E)}(t)$, asymmetry, $\gamma_a^{(E)}(t)$, and excess, $\gamma_e^{(E)}(t)$, of the reverberation envelope decrease, (we note that for the normal distribution, $\gamma_a(t) = 0$ and $\gamma_e(t) = 0$). On the other hand, in the absence of a quasi-deterministic component, i.e., when $q(t) = 0$, it follows from (46.34) that

*Or Rice law, after S. O. Rice, *Bell Syst. Tech. J.*, Vol. 24, 1945, p. 46, Eq. (3.10-11).

$$W(E,t)\sigma_F(t) = U\exp\left(-\frac{U^2}{2}\right)\left\{1 + \frac{\gamma_e(t)}{4!}\left(\frac{3U^4}{4} - 6U^2 + 6\right) + \right.$$
$$\left. + \left(\frac{\gamma_e(t)}{4!}\right)^2\left[\frac{3}{128}U^8 - \frac{3}{4}U^6 + \frac{27}{4}U^4 - 18U^2 + 9\right]\right\}. \quad (46.36)$$

This distribution (density) describes the effect of low scatterer density in the marine enviornment on the properties of the reverberation envelope in the absence of a quasideterministic component. An analysis of the relationship (46.36) shows that as the parameter $\gamma_e(t)$ increases -- that is, as the average scatterer density in the marine environment decreases, probability density $W(E,t)$ shifts to the left and deviates more and more from a normal distribution. In this case, as the value of $\gamma_e(t)$ grows, the coefficients of variation $\gamma_v^{(E)}(t)$, asymmetry $\gamma_a^{(E)}(t)$, and excess $\gamma_e^{(E)}(t)$, of this distribution increase. Finally, if $q(t) = 0$ and $\gamma_e(t) \to 0$, then as the limit we get the Rayleigh distribution

$$W(F,t)\sigma_F(t) = U\exp\left(-\frac{U^2}{2}\right), \quad (46.37)$$

which, as we know, is valid for the envelope of normal random processes.

Figure 10 shows graphs of the distribution curves $W(E,t)\sigma_F(t)$ plotted in accordance with the general relationship (46.34) for different values of the parameters $\gamma_e(t)$ and $q(t)$. It is evident from these graphs that as the coefficient of excess $[\gamma\ (t)]$ increases, the distribution density of the envelope shifts to the left and deviates more and more significantly from the Rayleigh and normal densities. As the level of the quasideterministic component increases -- that is, as the parameter $q(t)$ increases the probability density $W(E,t)$ shifts to the right and approximates a normal distribution, becoming symmetrical relative to the mean.

It is interesting to note that in terms of coherent scattering off of discrete irregularities, in the event that the average scatterer density -- that is, the parameter $n(t)$ -- increases, we can expect that the value of $q(t)$ will grow according to the law $q(t) \sim n^{1/2}(t)$. On the other hand, it follows from (46.11) that $\gamma_e(t) \sim 1/n(t)$. In this regard we are faced with the problem of finding a criterion by which to judge whether the number of scatterers is large or small. We can use the coefficients of variation, $\gamma_v(t)$,

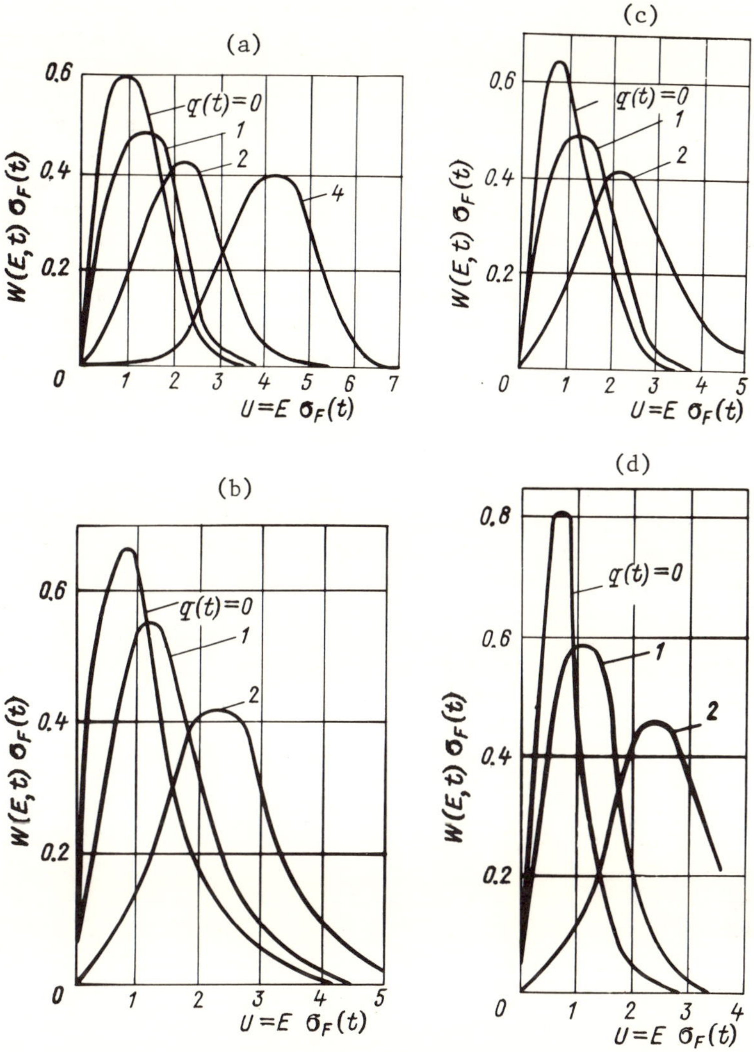

Figure 10. Probability densities of the envelope of reverberation at different values of the parameters $q(t)$ and $\gamma_e(t)$: *a* -- at $\gamma_e(t) = 0$; *b* -- at $\gamma_e(t) = 1$; *c* -- at $\gamma_e(t) = 2$; *d* -- at $\gamma_e = 4$.

and excess, $\gamma_e(t)$, as such a criterion. We can define $n(t)$, which characterizes the average density of scatterers in the medium, on the basis of these values. Thus, under certain conditions which promote an increase in the density of scatterers in the marine environment, the parameter $q(t)$ will increase and the parameter $\gamma_e(t)$ will decrease simultaneously. In other words, as the scatterer density increases, the probability density $W(E,t)$ quickly converges to a generalized Rayleigh distribution, (46.35), and this means that it approaches a normal distribution at large values of $q(t)$. If, however, the presence of a quasideterministic component in the reverberation process stems from the reflection of acoustic waves from individual scatterers larger than others, then the values of the parameters $\gamma_e(t)$ and $q(t)$ do not depend on each other and may associate randomly in expression (46.34) for $W(E,t)$.

On the basis of the analysis above we can conclude, in particular, that in conducting experimental research on the scattering properties of the water medium and its boundaries, in principle we can evaluate the average number of scattered signals generating a reverberation process, and the level of coherent scattering, by studying the types of one-dimensional distributions present for the reverberation signals and their envelope. In this case the smaller the expected values of the parameters $\gamma_e(t)$ and $q(t)$, the greater is the volume of experimental data which must be subjected to statistical treatment in measuring the probability distributions of reverberation signals. Some aspects of this problem are described in §65 and §66 following.*

---

*See also, [58,62]. (D.M.)

CHAPTER VII

PROBLEMS OF DETECTING ECHO SIGNALS IN THE PRESENCE OF INTERFERENCE

§47. Specific Features of Detecting Echo Signals in Sonar

The first task in processing sonar information is detecting the echo signal reflected from the object of detection. The following factors are significant in the execution of this task for sonar [13,14,22,36]:

1. the form of the echo signal may not match the form of the emitted signal in the general case;

2. the echo signal is received in the presence of a combination of reverberation and noise interference;

3. sonar conditions (underwater observation conditions) are extremely changeable in time and space.

These features of echo signal affect both the analysis and synthesis of sonar systems. Let us examine them in more detail.

In Chapter V we discussed different probability models of echo signals, in which case we devoted most of our attention to the so-called "characteristics of similarity" (e.g., cross-correlations) between emitted signals and echo signals, given different models of distortion for the latter. It is precisely these characteristics of similarity which define differences in the form of both types of signals [when they are described at the correlation (second-order second moment) level, of course]. When the coefficient of similarity is equal to one, then the task of detecting the echo signal reduces to detecting a signal of known form with random parameters (initial phase, amplitude, and so on). When the mathematical expectation of the coefficient of signal similarity equals zero, the task of detecting the echo signal reduces to detecting a signal of random form -- that is, a signal representing a segment of a random process. The structures of the optimum receivers are definite for both cases (on the assumption that interference is a random Gaussian process) [6,12,13,25,26,32]: A cross-correlation system is used to receive echo signals of known form, while a quadratic (i.e., "energy") detector is used to receive echo signals representing segments of random processes. However, both types of signals noted are not adequate input forms for the echo signal models we have examined here, since none of the models of practical interest fit them. This poses several problems specific to sonar, of which we will mention two:

(a) determining the detection characteristics of echo signals, given prescribed processing algorithms which take account of their probability models (the tasks of analysis);

(b) finding optimum detection systems, given that the form of the echo signal is known (the tasks of synthesis).

The second feature of echo signal detection specific to sonar is that reception goes on in the presence of two types of interference -- reverberation and background noise. Whereas noise interference can usually be considered "white noise" (within the frequency range of interest to us), reverberation interference (see Chapter VI) is so-called "colored noise," and its probability characteristics depend significantly on the form of the emitted signal. As a result, the following specific tasks arise in the detection of sonar echo signals:

(a) determining the detection characteristics of echo signals, given prescribed processing algorithms which take account of the effects of reverberation interference (the task of analysis);

(b) finding optimum detection systems when echo signals are received in the presence of noise and reverberation interference (the task of synthesis);

(c) selection of an efficient form of emitted signal (the task of synthesis).

Finally, the third feature of echo signal detection noted above is the changeability of underwater observation conditions. In this regard we lay primary emphasis on the variability of echo signal models and of the relationship between noise and reverberation interference. We find, in reality, that sonar systems are used in the presence of incomplete *a priori* information regarding probability properties of the echo signals and interference. This forces us to adapt the systems to changing external effects. Naturally, whenever probability models of signals and interference change in response to various conditions, adaptive sonar systems are more efficient than are "rigid" systems -- that is, systems with unalterable algorithms for information processing. An important aspect of this problem is that in selecting the characteristics of sonar systems we must make use not only of the *a priori* information regarding the probability properties of the underwater observation conditions, but also current information on the statistical characteristics of these conditions -- at least on the statistical characteristics of noise and reverberation interference under current sonar conditions.

In subsequent paragraphs of this chapter we will examine the problems of detecting sonar echo signals with a consideration for some of the features noted above. The results presented below are based on results published in [6,7,11-14,19,22,26].

§48. Probability Characteristics of Detection*

The quality of echo signal detection in sonar is evaluated by the probability characteristics of detection. These characteristics can be used to distinguish between properly detected echo signals and false alarms.

Let $X(t)$ be a process at the input of the sonar information processing system and Y be the effect at its

*The following is a well-known, important case of Bayes risk detection theory (cf. Secs. 19.1-19.3, for example; [12], for a general treatment). (D.M.)

output. Then if $\mathscr{P}_t$ is the algorithm of temporal processing,

$$Y = \mathscr{P}_t [X(t)]. \tag{48.1}$$

Let us examine Y as a random variable applicable to two situations, one of which corresponds to the presence of an echosignal and interference at the system's input (hypothesis H_1), while the other corresponds to the presence of interference alone at the input (hypothesis H_0). We designate by $W(Y|0)$ and $W(Y|1)$ the conditional probability densities of variable Y, corresponding to hypotheses H_0 and H_1. We will examine simple threshold detection of echo signals. Therefore, assuming that Y_0 is a certain threshold, we define the conditional probabilities of correct detection P_d and false alarms P_{fa} by the following relationships:

$$P_d = \int_{Y_0}^{\infty} W(Y|1)\,dY; \tag{48.2}$$

$$P_{fa} = \int_{Y_0}^{\infty} W(Y|0)\,dY. \tag{48.3}$$

Thus, we consider correct detection to be a situation in which $Y > Y_0$ and hypothesis H_1 holds, while a false alarm is a situation in which $Y > Y_0$ and hypothesis H_0 holds. Equation (48.3) permits us to determine the value of threshold Y_0 given a certain probability density $W(Y|0)$ and the probability of false alarms:

$$Y_0 = \varphi(P_{fa}), \tag{48.4}$$

where φ is a function whose form depends on $W(Y|0)$. Then, substituting (48.4) into (48.2) we have

$$P_d = \int_{\varphi(P_{fa})}^{\infty} W(Y|1)\,dY. \tag{48.5}$$

Equation (48.5), which relates probabilities P_d and P_{fa}, the properties of the input process $X(t)$, and the characteristics of operator $\mathscr{P}_t$, defines the detection characteristics of echo signals.

Several criteria of effectiveness in solving such problems are known in detection theory. The following is one of the more general:

$$R = \alpha_1 P (1 - P_d) + \alpha_2 (1 - P) P_{fa}, \quad (48.6)$$

in which case R is the average risk of making incorrect decisions (admission of a signal and false alarms), P is the *a priori* probability for the presence of a signal, and α_1 and α_2 are weight factors defining the "cost" of incorrect decisions. If, however, we cannot assign values for α_1, α_2, and P or if we do not have adequate grounds for selecting them, then it may be convenient to use the conditional probability of correct detection P_d given a fixed, e.g.,

$$P_{fa} = \text{const.} \quad (48.7)$$

as the criterion. In a special sense this (Neyman-Pearson) criterion follows from (48.6) and is typical of the situation in active sonar.

The detection characteristic, $P_d = F(P_{fa})$, of sonar echo signals is most frequently set up as a collection of curves, each of which shows detection probability as a function of parameters defining the properties of the process at the processing system's input (such a parameter can be, for example, the signal to interference ratio). In this case the probability of false alarms is held constant.

§49. Typical Probability Distributions of the Output Effect

As had been noted in §48, we observe a random variable Y at the output of the detection system. This variable is compared with a threshold Y_0. Below we will examine typical probability densities $W(Y|0)$ and $W(Y|1)$ which correspond to the presence of interference only and to a mixture of echo signal and interference at the output.

(i) We now assume that $W(Y|0)$ and $W(Y|1)$ are Gaussian probability densities of the random variable Y. Such probability distributions are typical of information processing systems in which integration is performed at their outputs over a time interval significantly exceeding that of interference correlation. In this case even if the distribution of the process is not itself Gaussian at the integrator's input, this distribution becomes "normalized" at the output in view of the central limit theorem of probability theory. Considering this, we have

$$W(Y \mid 0) = \frac{1}{\sqrt{2\pi d(0)}} \exp\left[-\frac{Y^2}{2d(0)}\right], \tag{49.1}$$

$$W(Y \mid 1) = \frac{1}{\sqrt{2\pi d(1)}} \exp\left[-\frac{[Y - m(1)]^2}{2d(1)}\right], \tag{49.2}$$

where $d(0)$ and $d(1)$ are the variances of Y corresponding to the absence or presence of an echo signal, and $m(1)$ is the mathematical expectation of Y in the presence of an echo signal. In this case it is assumed that $m(0) = 0$ -- that is, that in the absence of an echo signal the mathematical expectation of the output effect is zero.

Let us define the probability of false alarms P_{fa}. In accordance with (48.3) and (49.1) we have

$$P_{\text{fa}} = \frac{1}{\sqrt{2\pi d(0)}} \int_{Y_0}^{\infty} \exp\left(-\frac{Y^2}{2d(0)}\right) dY. \tag{49.3}$$

We use one definition of the probability integral:

$$\Phi(x) \equiv \frac{2}{\sqrt{2\pi}} \int_0^x \exp\left(-\frac{t^2}{2}\right) dt. \tag{49.4}$$

Then, from relationships (49.3) and (49.4) we find that

$$P_{\text{fa}} = \frac{1}{2}\left[1 - \Phi\left(\frac{Y_0}{\sqrt{d(0)}}\right)\right]. \tag{49.5}$$

We next introduce the function $\Phi^{-1}(U)$, the inverse of $\Phi(x)$ -- that is, such a function that if $U = \Phi(x)$, then $x = \Phi^{-1}(U)$. In this case, from (49.5), we have

$$Y_0 = \sqrt{d(0)}\, \Phi^{-1}(1 - 2P_{\text{fa}}). \tag{49.6}$$

We can similarly find the detection probability P_{d} on the basis of the relationship (48.2) using (49.2):

$$P_{\text{d}} = \frac{1}{\sqrt{2\pi d(1)}} \int_{Y_0}^{\infty} \exp\left[-\frac{(Y - m(1))^2}{2d(1)}\right] dY. \tag{49.7}$$

From (49.4) and (49.7) we get

$$P_{\rm d}=\frac{1}{2}\left[1+\Phi\left(\frac{m(1)-Y_0}{\sqrt{d(1)}}\right)\right],\quad m(1)\geqslant Y_0. \tag{49.8}$$

Now, it is not difficult to get the detection characteristic of echo signals. Substituting (49.6) into (49.8) we find that

$$P_{\rm d}=\frac{1}{2}\left\{1+\Phi\left[\frac{m(1)-\sqrt{d(0)}\Phi^{-1}(1-2P_{\rm fa})}{\sqrt{d(1)}}\right]\right\}. \tag{49.9}$$

(ii) The next typical distribution of random variables follows Rayleigh's law. This law applies in the event that the information processing system is set up as a high-frequency filter, linear detector, and integrator, and where the echo signal envelope fluctuates in accordance with Rayleigh's law. Such a processing system is widely employed in sonar. In this case we have

$$W(Y\mid 0)=\frac{Y}{d_X(0)}\exp\left(-\frac{Y^2}{2d_X(0)}\right),\ Y\geqslant 0, \tag{49.10}$$

$$W(Y\mid 1)=\frac{Y}{d_X(1)}\exp\left(-\frac{Y^2}{2d_X(1)}\right),\ Y\geqslant 0, \tag{49.11}$$

where $d_X(0)$ and $d_X(1)$ are the variances of the random processing system's output. We note that inasmuch as the envelope of the Gaussian process conforms to Rayleigh's law, these variances pertain precisely to this process. Thus, in the case being examined here we assume that both interference and the echo signal and interference mixture at the processing system's input have a Gaussian distribution.

In accordance with (48.3) and (49.10) the probability of false alarms $P_{\rm fa}$ is defined by the relationship

$$P_{\rm fa}=\exp\left(-\frac{Y_0^2}{2d_X(0)}\right), \tag{49.12}$$

whence

$$Y_0=\sqrt{2d_X(0)\ln\frac{1}{P_{\rm fa}}}. \tag{49.13}$$

On the basis of (48.2) and (49.11) we find that the detection probability P_d is equal to

$$P_d = \exp\left(-\frac{Y_0^2}{2d_X(1)}\right). \tag{49.14}$$

Substituting the threshold value Y_0, as defined by (49.13), into (49.14), we find the detection characteristic:

$$P_d = \exp\left(\frac{d_X(0)\ln P_{fa}}{d_X(1)}\right). \tag{49.15}$$

(iii) The next typical probability distributions for variable Y are Rayleigh's law for interference and a generalized Rayleigh's law (Rice's distribution)* for an echo signal and interference combination at the output of the information processing system. Such distributions are observed at the output of a high-frequency filter, linear detector, integrator system in the event that the echo signal has a component possessing a constant level, and where interference at the system's output has a Gaussian distribution. In this case we have

$$W(Y\mid 0) = \frac{Y}{d_X(0)}\exp\left(-\frac{Y^2}{2d_X(0)}\right), \quad Y \geqslant 0; \tag{49.16}$$

$$W(Y\mid 1) = \frac{Y}{d_X(1)}\exp\left(-\frac{Y^2+S_0^2}{2d_X(1)}\right) I_0\left(\frac{YS_0}{d_X(1)}\right), \quad Y \geqslant 0, \tag{49.17}$$

where S_0 is the echo signal amplitude, while $d_X(0)$ and $d_X(1)$ are interpreted as variances in the fluctuation of the input process $X(t)$ in the absence and in presence of an echo signal, respectively.

We can find the probability of false alarms P_{fa} using relationships (48.3) and (49.16):

$$P_{fa} = \exp\left(-\frac{Y_0^2}{2d_X(0)}\right), \tag{49.18}$$

whence

$$Y_0 = \sqrt{2d_X(0)\ln\frac{1}{P_{fa}}}. \tag{49.19}$$

*See footnote on page 143. (D.M.)

According to (48.2) and (49.17) the detection probability is now given by

$$P_d = \frac{1}{d_X(1)} \int_{Y_0}^{\infty} Y \exp\left(-\frac{Y^2 + S_0^2}{2d_X(1)}\right) I_0\left(\frac{YS_0}{d_X(1)}\right) dY. \qquad (49.20)$$

Using the tabulated function

$$Q(a, b) \equiv \int_a^{\infty} y \exp\left(-\frac{y^2 + b^2}{2}\right) I_0(by)\, dy \qquad (49.21)$$

(Q-function),* on the basis of (49.19)-(49.21) we find P_d to be

$$P_d = Q\left(\sqrt{2\frac{d_X(0)}{d_X(1)} \ln \frac{1}{P_{fa}}},\ \frac{S_0}{\sqrt{d_X(1)}}\right). \qquad (49.22)$$

The distributions presented above represent most of the typical probability distributions for the random variable Y observed at the output of a sonar information processing system. We will subsequently go on to an examination of echo signal detection characteristics relevant to specific (i.e., usually suboptimum) information processing systems.

§50. Incoherent Detection of Simple Echo Signals with a Linear Detector

Let us examine a receiver set up as a high-frequency filter, linear detector, and integrator. We will study two cases, namely the incoherent reception of an echo signal with constant amplitude and one with an amplitude that fluctuates in accordance with Rayleigh's law. In this case we will assume that the high-frequency filter is optimum (matched to the echo signal), the form of the echo signal is known, and the detector's time constant is close to the echo signal duration. When simple signals are emitted, such assumptions follow naturally, and they pertain to a wide class of underwater observation conditions. If the echo signal has constant amplitude S_0 and sinusoidal occu-

*See, for example, [45], p. 114, and references. (D.M.)

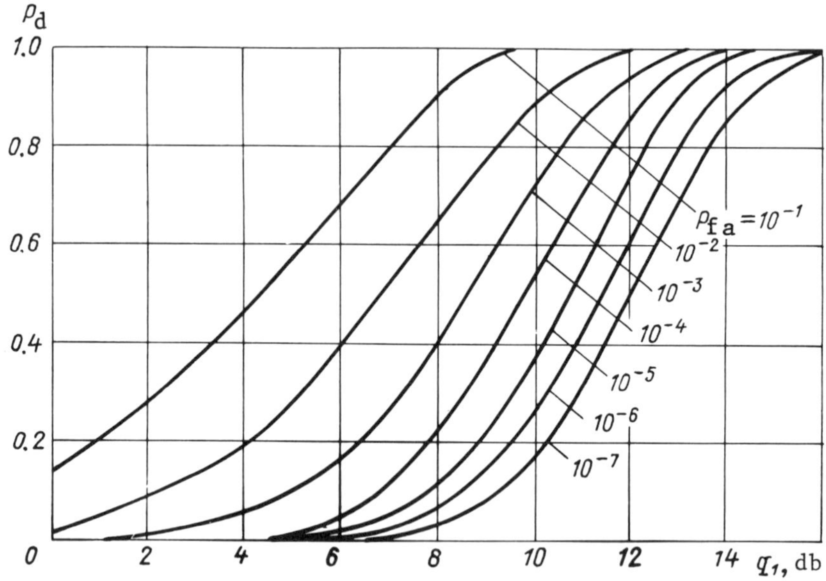

Figure 11. Detection characteristics of simple echo signals with constant amplitude, using linear detection.*

pation, then the relation (49.22) is valid for the detection probability P_d at $d_X(0) = d_X(1) = d_X$ -- that is,

$$P_d = Q\left(\sqrt{2\ln\frac{1}{P_{fa}}},\ \sqrt{2q_1}\right), \tag{50.1}$$

where

$$q_1 = \frac{S_0^2}{2d_X} \tag{50.2}$$

is the signal/interference ratio. We note that in this case d_X is the variance of the interference computed for the band of the high-frequency filter. Given optimum filtration, the filter's effective pass band is $\Delta\omega_{ef} \simeq 2\pi/T_{ef}$, where T_{ef} is the signal's effective duration and $d_X = G\Delta\omega_{ef}$, where G is the power spectrum of interference about the fil-

*I.e., incoherent envelope detection. (D.M.)

ter's central frequency. In this case it is obvious that

$$q_1 = \frac{E_{ef}}{2\pi G}, \tag{50.3}$$

where $E_{ef} = S_0^2 T_{ef}/2$ is the echo signal's effective energy.

Detection characteristics corresponding to reception of such a simple echo signal with constant level are presented in Figure 11.

Let us now examine linear detection of a fluctuating echo signal. If the echo signal amplitude fluctuates in accordance with Rayleigh's law, and if the signal has sinusoidal occupation, then relationship (49.15) is valid for the detection probability -- that is,

$$P_d = \exp\left(\frac{\ln P_{fa}}{1+q_2}\right), \tag{50.4}$$

where

$$q_2 = \frac{d_X(1) - d_X(0)}{d_X(0)} = \frac{d_S}{d_X(0)} \tag{50.5}$$

is the ratio between the variance of the echo signal fluctuation d_S and the variance of the interference $d_X(0)$.

Figure 12 shows the detection characteristics corresponding to this case. We note that in both cases examined above the values of the signal/interference ratio q differ quantitatively somewhat: When an echo signal with constant amplitude is received -- formula (50.2) -- the value of q_1 is computed on the basis of echo signal amplitude S_0 and the variance of interference d_X. On the other hand, when an echo signal with fluctuating amplitude is received -- formula (50.5) -- the value of q_2 is computed on the basis of the variances of the echo signal d_S and interference d_X. Of course, this difference should be taken into account when the characteristics shown in Figures 11 and 12 are used.

§51. Incoherent Detection of a Noiselike Echo Signal with Quadratic Detection and Smoothing ("Energy Detection")

In this case the information processing system consists of the following series of components -- high-

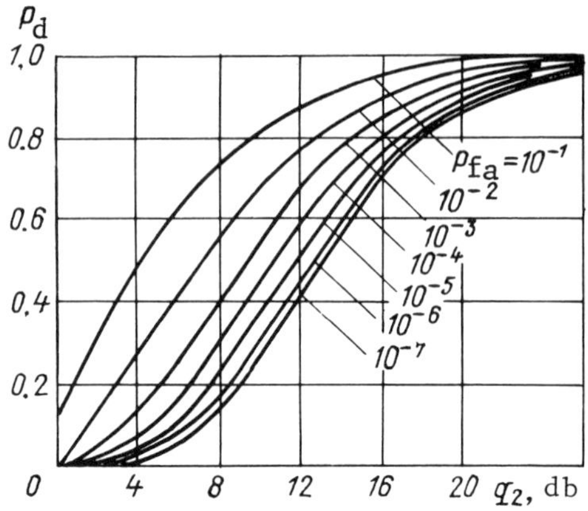

Figure 12. Detection characteristics of a simple echo signal with fluctuating amplitude, using linear detection

frequency filter, quadratic detector, integrator. Let the process

$$x(t) = S(t) + N(t), \tag{51.1}$$

be observed at the output of the high-frequency filter, where $S(t)$ is the echo signal and $N(t)$ is the interference. Then the random variable

$$Y = \frac{1}{T}\int_0^T x^2(t)\,dt \tag{51.2}$$

is formed at the output of the processing system, where T is the echo signal duration. We are interested in the two situations in which

$$Y(0) = \frac{1}{T}\int_0^T N^2(t)\,dt - m; \tag{51.3}$$

$$Y(1) = \frac{1}{T}\int_0^T [S(t) + N(t)]^2\,dt - m, \tag{51.4}$$

in which case $Y(0)$ corresponds to the absence of an echo signal and $Y(1)$ corresponds to the presence of an echo signal at the system's input, and where $m = \langle N^2(t)\rangle$ is the mathematical expectation of the square of the noise interference (i.e., the noise intensity).

In this case we assume the echo signal and interference to be wideband processes. Therefore, we can consider the probability distributions $Y(0)$ and $Y(1)$ to be Gaussian at the output of the smoothing device. Then the relation (49.9), containing the following parameters, is valid for the detection characteristic:

(a) *mathematical expectation* of the random variable (51.4)

$$m(1) = \langle Y(1) \rangle; \tag{51.5}$$

(b) *variance* of the random variable (51.3)

$$d(0) = \langle Y^2(0) \rangle; \tag{51.6}$$

(c) *variance* of the random variable (51.4)

$$d(1) = \langle Y^2(1) \rangle - \langle Y(1) \rangle^2. \tag{51.7}$$

Let us define these parameters.

Substituting (51.4) into (51.5), we find that

$$\begin{aligned} m(1) &= \frac{1}{T}\int_0^T \langle [S(t) + N(t)]^2 \rangle\, dt - m \\ &= \frac{1}{T}\int_0^T \langle S^2(t) \rangle\, dt + \frac{2}{T}\int_0^T \langle S(t)\, N(t) \rangle dt + \frac{1}{T}\int_0^T \langle N^2(t) \rangle\, dt - m. \end{aligned} \tag{51.8}$$

We designate by I_S and I_N the average intensities of the echo signal and interference, respectively:

$$I_S = \langle S^2(t) \rangle; \tag{51.9}$$

$$I_N = \langle N^2(t) \rangle = m \tag{51.10}$$

and take account of the fact that the echo signal and interference are uncorrelated -- that is,

$$\langle S(t)\, N(t) \rangle = 0. \tag{51.11}$$

Then, on the basis of (51.8)-(51.11) we get

$$m(1) = I_S. \tag{51.12}$$

Next, let us find the variance $d(0)$. Substituting (51.3) into (51.6), we have

$$d(0) = \left\langle \left[\frac{1}{T} \int_0^T N^2(t)\, dt - m \right]^2 \right\rangle$$

$$= \frac{1}{T^2} \int_0^T \int_0^T \langle N^2(t') N^2(t'') \rangle\, dt'\, dt'' - \frac{2m}{T} \int_0^T \langle N^2(t) \rangle\, dt + m^2. \tag{51.13}$$

Assuming that the interference $N(t)$ is Gaussian,* we can write:

$$\langle N^2(t') N^2(t'') \rangle = I_N^2 + 2I_N^2 R_N^2(t' - t''), \tag{51.14}$$

where $R_N(t' - t'')$ is the normalized autocorrelation function of the noise interference.

Using relationships (51.10)-(51.14) we get the following representation for the variance:

$$d(0) = \frac{4I_N^2}{T} \int_0^T \left(1 - \frac{\tau}{T}\right) R_N^2(\tau)\, d\tau. \tag{51.15}$$

The variance $d(1)$ can be found by substituting (51.4) into (51.7). Considering (51.9) and (51.10), we have

$$d(1) = \left\langle \left\{ \frac{1}{T} \int_0^T [S(t) + N(t)]^2\, dt - \left\langle \frac{1}{T} \int_0^T [S(t) + N(t)]^2\, dt \right\rangle \right\}^2 \right\rangle =$$

$$= \frac{1}{T^2} \int_0^T \int_0^T \langle S^2(t') S^2(t'') \rangle\, dt'\, dt'' + \frac{4}{T^2} \int_0^T \int_0^T \langle S(t') N(t') S(t'') N(t'') \rangle + dt'\, dt'' +$$

$$+ \frac{1}{T^2} \int_0^T \int_0^T \langle N^2(t') N^2(t'') \rangle\, dt'\, dt'' + \frac{4}{T^2} \int_0^T \int_0^T \langle S^2(t') S(t'') N(t'') \rangle\, dt'\, dt'' +$$

$$+ \frac{2}{T^2} \int_0^T \int_0^T \langle S^2(t') N^2(t'') \rangle\, dt'\, dt'' + \frac{4}{T^2} \int_0^T \int_0^T \langle S(t') N(t') N^2(t'') \rangle\, dt'\, dt'' - (I_S + I_N)^2. \tag{51.16}$$

---

*If $x(t)$ is a Gaussian stationary random process, then the following relationship is valid [12]; Sec. 7.3-2:

$$\langle x(t_1)\, x(t_2)\, x(t_3)\, x(t_4) \rangle = \langle x(t_1)\, x(t_2) \rangle \langle x(t_3)\, x(t_4) \rangle + \langle x(t_1)\, x(t_3) \rangle \langle x(t_2)\, x(t_4) \rangle + \langle x(t_1)\, x(t_4) \rangle \langle x(t_2)\, x(t_3) \rangle.$$

We next assume that both the echo signal $S(t)$ and the interference $N(t)$ are Gaussian random processes with normalized autocorrelation functions $R_S(\tau)$ and $R_N(\tau)$ respectively. Then

$$\left.\begin{aligned} \langle S^2(t')\, S^2(t'')\rangle &= I_S^2 + 2I_S^2 R_S^2(t'-t''); \\ \langle S(t')\, N(t')\, S(t'')\, N(t'')\rangle &= I_S I_N R_S(t'-t'')\, R_N(t'-t''); \\ \langle N^2(t')\, N^2(t'')\rangle &= I_N^2 + 2I_N^2 R_N^2(t'-t''); \\ \langle S^2(t')\, S(t'')\, N(t'')\rangle &= 0; \\ \langle S^2(t')\, N^2(t'')\rangle &= I_S I_N; \\ \langle S(t')\, N(t')\, N^2(t'')\rangle &= 0. \end{aligned}\right\} \qquad (51.17)$$

Substituting (51.17) into (51.16) and performing a simple transformation, we find that

$$d(1) = \frac{4I_S^2}{T}\int_0^T \left(1-\frac{\tau}{T}\right) R_S^2(\tau)\, d\tau + \frac{8I_S I_N}{T}\int_0^T \left(1-\frac{\tau}{T}\right) R_S(\tau)\, R_N(\tau)\, d\tau + \frac{4I_N^2}{T}\int_0^T \left(1-\frac{\tau}{T}\right) R_N^2(\tau)\, d\tau. \qquad (51.18)$$

It is not difficult to see that we get (51.15) from (51.18) if we assume $I_S = 0$ -- that is, if we consider that there is no echo signal at the input.

As an example, we will examine a case in which the autocorrelation functions of the echo signal and interference have the same form and are defined by the relationship

$$R(\tau) = R_S(\tau) = R_N(\tau) = \exp(-2\,\Delta F|\tau|)\cos(2\pi f_0 \tau), \qquad (51.19)$$

where ΔF is the effective bandwidth and f_0 is the spectrum's central frequency. If the conditions

$$\Delta F T \gg 1, \ \Delta F \ll f_0, \qquad (51.20)$$

are satisfied, then we can assume that

$$\frac{1}{T}\int_0^T\left(1-\frac{\tau}{T}\right)R^2(\tau)\,d\tau=\frac{1}{8\,\Delta FT}. \quad (51.21)$$

Next, we introduce the parameter

$$q_3=\frac{I_S}{I_N}, \quad (51.22)$$

which defines the ratio between the average intensities of the signal and interference in the high-frequency filter band ΔF.* Then from $m(1)$, $d(0)$, and $d(1)$ we get the following on the basis of (51.12), (51.15), (51.18), (51.19), (51.21), and (51.22):

$$m(1)=I_N q_3; \quad (51.23)$$

$$d(0)=\frac{I_N^2}{2\,\Delta FT}; \quad (51.24)$$

$$d(1)=\frac{I_N^2}{\Delta FT}\left(\frac{q_3^2}{2}+q_3+\frac{1}{2}\right). \quad (51.25)$$

Now we can write the expression for the echo signal detection characteristic which corresponds to the case examined here. Using (49.9) and (51.23)-(51.25) we have

$$P_{\text{d}}=\frac{1}{2}\left\{1+\Phi\left[\frac{q_3\sqrt{2\,\Delta FT}-\Phi^{-1}\left(1-2P_{\text{fa}}\right)}{q_3+1}\right]\right\}. \quad (51.26)$$

Figure 13 shows these detection characterisitcs plotted for two values of the parameters ΔFT and different values of the probability of false alarm P_{fa}.

---

*It should be kept in mind that the parameter q_3 does not agree, by its definition, with parameters q_1 and q_2 (see §50). In particular, we can show that $q_3\approx q_1/\Delta FT$, $q_3\approx q_2/\Delta FT$, in which case q_1 and q_2 pertain to simple signals, while q_3 pertains to a complex noiselike signal.

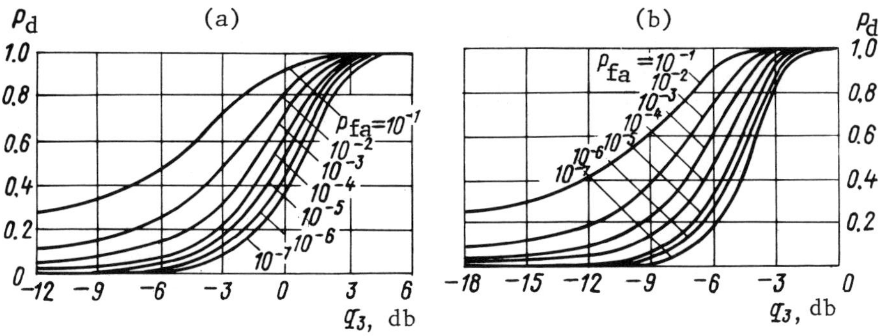

Figure 13. Detection characteristics for a noiselike echo signal using quadratic detection: (a) at $\Delta FT = 10$; (b) at $\Delta FT = 10^2$.

§52. Coherent Detection of a Noiselike Echo Signal Using Cross-Correlation Processing

When cross-correlation (i.e., "matched filter"*) processing of input information is employed, the receiver has the following structure -- input process, high-frequency filter, and expected signal cross-multiplier, integrator. In other words if the process

$$x(t) = S(t) + N(t), \tag{52.1}$$

is present at the output of the high-frequency filter, where $S(t)$ is the echo signal and $N(t)$ is the interference, then the random variable

$$Y = \frac{1}{T}\int_0^T x(t)C(t)\,dt, \tag{52.2}$$

is formed at the output of the processing system, where $C(t)$ is the emitted signal and T is its duration. On the basis of (52.1) and (52.2) the two random variables $Y(0)$ and $Y(1)$, corresponding to the absence and presence of an echo signal at the input of the information processing system, are defined by

$$Y(0) = \frac{1}{T}\int_0^T N(t)C(t)\,dt; \tag{52.3}$$

*See [38]. (D.M.)

$$Y(1)=\frac{1}{T}\int_0^T [S(t)+N(t)]\,C(t)\,dt. \qquad (52.4)$$

As was true in the case of the incoherent processing a noiselike signal using quadratic detection (see §51), we assume here that the emitted signal, echo signal, and interference are wideband processes. Therefore, the probability distributions of random variables $Y(0)$ and $Y(1)$ are assumed to be Gaussian, and we must make use of the relationship (49.9) to set up the detection characteristics. Let us define the distribution parameters of (49.9).

The mathematical expectation

$$m(1)\equiv\langle Y(1)\rangle \qquad (52.5)$$

is specified by substituting (52.4) into (52.5):

$$m(1)=\left\langle \frac{1}{T}\int_0^T [S(t)+N(t)]\,C(t)\,dt\right\rangle$$
$$=\frac{1}{T}\int_0^T \langle S(t)\,C(t)\rangle\,dt+\frac{1}{T}\int_0^T \langle N(t)\,C(t)\rangle dt. \qquad (52.6)$$

We define the cross-correlation function of the emitted signal and echo signal by*

$$B_{CS}(t''-t')\equiv\langle S(t')\,C(t'')\rangle=\sqrt{I_C I_S}\,R_{CS}(t''-t')$$
$$=\sqrt{I_C I_S}\,R_{CS}R_C(t''-t'), \qquad (52.7)$$

where

$$I_C=\langle C^2(t)\rangle \qquad (52.8)$$

is the average intensity of the emitted signal,

$$I_S=\langle S^2(t)\rangle \qquad (52.9)$$

is the average intensity of the echo signal, $R_{CS}(t''-t')$ is the normalized cross-correlation function of the emitted signal and echo signal, $R_{CS}=R_{CS}(t''-t')|_{t''=t'}$, and

*In eq. (52.7) it is assumed that $R_c(\tau)=R_s(\tau)$, cf. (52.17)*ff*. (D.M.)

$R_C(t'' - t')$ is the normalized autocorrelation function of the emitted signal. We will postulate that the emitted signal and interference are uncorrelated -- that is,

$$\langle N(t')\, C(t'')\rangle = 0. \tag{52.10}$$

On the basis of (52.6), (52.7), (52.9), and (52.10), for $m(1)$ we get

$$m(1) = \sqrt{I_C I_S}\, R_{CS}\,. \tag{52.11}$$

Next, we define the variance

$$d(1) \equiv \langle Y^2(1)\rangle - \langle Y(1)\rangle^2 \tag{52.12}$$

of the random variable $Y(1)$. Substituting (52.4) into (52.12), we have

$$\begin{aligned} d(1) = {} & \left\langle \frac{1}{T^2} \int_0^T \int_0^T [S(t') + N(t')]\,[S(t') + N(t'')]\, C(t')\, C(t'')\, dt'\, dt'' \right\rangle - \\ & - \left\langle \frac{1}{T} \int_0^T [S(t) + N(t)]\, C(t)\, dt \right\rangle^2 = \\ & = \frac{1}{T^2} \int_0^T \int_0^T \langle S(t')\, S(t'')\, C(t')\, C(t'')\rangle\, dt'\, dt'' + \\ & + \frac{2}{T^2} \int_0^T \int_0^T \langle S(t')\, N(t'')\, C(t')\, C(t'')\rangle\, dt'\, dt'' + \\ & + \frac{1}{T^2} \int_0^T \int_0^T \langle N(t')\, N(t'')\, C(t')\, C(t'')\rangle\, dt'\, dt'' - I_C I_S R_{CS}^2. \end{aligned} \tag{52.13}$$

Assuming that the emitted noiselike signal $C(t)$, the echo signal $S(t)$, and interference $N(t)$ all have Gaussian probability distributions, and designating by $R_C(\tau)$, $R_S(\tau)$, and $R_N(\tau)$ the normalized correlation functions for the processes indicated above, and considering that there is no correlation between the emitted signal and echo signal and the interference, we can write the following equalities:

$$\left.\begin{aligned}&\langle S(t')\,S(t'')\,C(t')\,C(t'')\rangle = I_C I_S R_{CS}^2 + \\ &+ I_C I_S R_{CS}^2 R_C^2(t''-t') + I_C I_S R_C(t''-t')\,R_S(t''-t'); \\ &\langle S(t')\,N(t'')\,C(t')\,C(t'')\rangle = 0; \\ &\langle N(t')\,N(t'')\,C(t')\,C(t'')\rangle = I_C I_N R_C(t''-t')\,R_N(t''-t').\end{aligned}\right\} \tag{52.14}$$

Substituting (52.14) into (52.13) and making the relevant transformations, we get

$$\begin{aligned} d(1) = {} & \frac{2I_C I_S}{T}\int_0^T \left(1-\frac{\tau}{T}\right) R_C(\tau)\,R_S(\tau)\,d\tau + \\ & + \frac{2I_C I_S R_{CS}^2}{T}\int_0^T \left(1-\frac{\tau}{T}\right) R_C^2(\tau)\,d\tau + \\ & + \frac{2I_C I_N}{T}\int_0^T \left(1-\frac{\tau}{T}\right) R_C(\tau)\,R_N(\tau)\,d\tau. \end{aligned} \tag{52.15}$$

Assuming in (52.15) that $I_S = 0$, we find the variance $d(0)$ in the absence of an echo signal to be

$$d(0) = \frac{2I_C I_N}{T}\int_0^T \left(1-\frac{\tau}{T}\right) R_C(\tau)\,R_N(\tau)\,d\tau. \tag{52.16}$$

Let us examine an example in which

$$R(\tau) = R_C(\tau) = R_N(\tau) = R_S(\tau) = \exp(-2\Delta F\,|\tau|)\cos(2\pi f_0\tau), \tag{52.17}$$

where ΔF and f_0 are the effective width and central frequency of the spectrum respectively of the emitted signal, echo signal, and the interference. Given the conditions

$$\Delta FT \gg 1, \qquad \Delta F \ll f_0, \tag{52.18}$$

and on the basis of (52.17) we have

$$\frac{1}{T}\int_0^T \left(1-\frac{\tau}{T}\right) R(\tau)\,d\tau = \frac{1}{8\,\Delta FT}. \tag{52.19}$$

Then, from (52.11), (52.15)-(52.17), and (52.19), we get the following values of the parameters of interest to us for these distributions of the random variable Y:

$$m(1) = \sqrt{I_C I_S}\, R_{CS}; \tag{52.20}$$

$$d(0) = \frac{I_C I_N}{4\,\Delta FT}; \tag{52.21}$$

$$d(1) = \frac{I_C I_S}{4\,\Delta FT} + \frac{I_C I_S R_{CS}}{4\,\Delta FT} + \frac{I_C I_N}{4\,\Delta FT}. \tag{52.22}$$

The detection characteristic for the echo signal can be defined by substituting (52.20)-(52.22) into (49.9). Designating by q_3 the ratio between the rms intensities of the signal and interference:

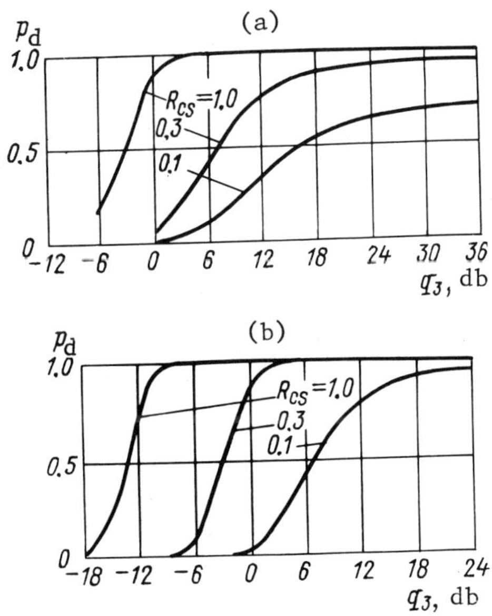

Figure 14. Detection characteristics of a noiselike echo signal using cross-correlation processing: (a) at $\Delta FT = 10$, $P_{fa} = 10^{-5}$; (b) at $\Delta FT = 10^2$, $P_{fa} = 10^{-5}$.

$$q_3 \equiv \sqrt{\frac{I_S}{I_N}}; \qquad (52.23)^*$$

we then find that

$$P_{\mathrm{d}} = \frac{1}{2}\left\{1 + \Phi\left[\frac{2q_3R_{CS}\sqrt{\Delta FT} - \Phi^{-1}(1 - 2P_{\mathrm{fa}})}{\sqrt{q_3^2 + q_3^2R_{CS} + 1}}\right]\right\}. \qquad (52.24)$$

Figure 14 shows the detection characteristics corresponding to the case examined here, plotted for different values of the parameters ΔFT and R_{CS} for a fixed level of P_{fa}.

§53. The Effect of Reverberation Interference on Echo Signal Detection

In examining the echo signal detection characteristics in the preceding paragraphs of this chapter we had not specified the type of interference to which function $N(t)$ is related and the specific conditions it satisfies. It had only been noted that $N(t)$ is a Gaussian random process with a spectrum matching that of the echo signal. This assumption was natural upon considering that the characteristics of the echo signal and interference were examined at the output of high-frequency filters which matched the echo signals. In this case parameter q was treated as the ratio between the average intensities of the echo signal and interference at the output of such a filter, and it had to be established that the given probability of detecting the echo signal at fixed values of the remaining parameters (probability of false alarm, and sometimes other parameters, as well) could be maintained. With such an approach, as a matter of course, we did not have to specify the form of interference acting on the detection system, because whether it be noise interference, reverberation interference, or their combination, the value of the parameter q necessary for detection, given the assumed hypotheses on the properties of the echo signal, interference, and the information processing algorithm, would still have to be the same.

*Equation (52.24), and hence Fig. 14, is correct provided q_3 is now redefined according to (52.23). (D.M.)

The situation is different when we formulate the task of analyzing the detection characteristics of echo signals as follows: The information processing algorithm and the interference models (noise and reverberation in the general case) are given; we must determine the effect of the type of emitted signal on the echo signal detection characteristics. We note that in this approach we are discussing the task of analysis, and not synthesis, in which we find an optimum, in a certain sense, type of emitted signal and algorithm for processing sonar information. Some of the problems of synthesis had been examined in [14,19,22,35].

Let us make this task concrete. Let there be a process with complex envelope

$$Z_X(t) = Z_S(t) + Z_F(t) + Z_N(t), \qquad (53.1)$$

at the input of the processing system, where $Z_S(t)$, $Z_F(t)$, and $Z_N(t)$ are the complex envelopes of the echo signal, reverberation, and noise interference respectively.

The processing system will be a correlator with process $Z_X(t)$ being fed into one input and process $Z^*_{C\Omega\tau}(t)$ being fed into the second, in which case

$$Z_{C\Omega\tau}(t) = Z_C(t-\tau) \exp[j\Omega(t-\tau)] \qquad (53.2)$$

is the complex envelope of the emitted signal $Z_C(t)$ shifted in time by amount τ and in frequency by amount Ω.

As has been indicated in §28, we note that movement of the object of detection causes not only a change in the central frequency of the spectrum by the amount Ω, corresponding to the relative rate of movement of the object, but also changes in the form of the echo signal as compared to that of the emitted signal. This effect is particularly significant when complex signals are used as the emitted signals. However, we will assume either that these changes are compensated, since we know them for each value of the object's rate of movement, or that we are using signals which are not changed by these distortions (for example, frequency-modulation obeying a hyperbolic law). We will examine random variable Y at the information processing system's output at a time when the echo signal exists.

Considering the above, we have the following random variables at the output of this cross-correlation system:

In the absence of an echo signal we have

$$Y(0) = \int_{-\infty}^{\infty} Z_F(t) Z_C^*(t) \exp(-j\Omega t)\, dt + \int_{-\infty}^{\infty} Z_N(t) Z_C^*(t) \exp(-j\Omega t)\, dt; \tag{53.3}$$

in the presence of an echo signal

$$Y(1) = \int_{-\infty}^{\infty} Z_S(t) Z_C^*(t) \exp(j\Omega t)\, dt + \int_{-\infty}^{\infty} Z_F(t) Z_C^*(t) \exp(j\Omega t)\, dt + \int_{-\infty}^{\infty} Z_N(t) Z_C^*(t) \exp(j\Omega t)\, dt. \tag{53.4}$$

If we postulate that the random variables $Y(0)$ and $Y(1)$ have a Gaussian probability distribution, then the echo signal detection characteristic is defined by (49.9). To define this characteristic we must then compute parameters $m(1)$, $d(0)$, and $d(1)$ for the random variables (53.3) and (53.4)

We introduce some additional assumptions pertaining to the properties of the signals and interference under examination.

We will consider that the emitted signals are deterministic and that the emitted signal and the echo signal are not correlated with the interference. We also assume that the different types of interference are not correlated among each other -- that is,

$$\left.\begin{aligned} \langle Z_C(t) Z_F(t)\rangle = 0;^{*} \quad \langle Z_C(t) Z_N(t)\rangle = 0; \\ \langle Z_S(t) Z_F(t)\rangle = 0;^{*} \quad \langle Z_S(t) Z_N(t)\rangle = 0; \\ \langle Z_F(t) Z_N(t)\rangle = 0. \end{aligned}\right\} \tag{53.5}$$

---

*This implies, also, that there is no coherent scatter component in the reverberation. (D.M.)

In addition, we present the echo signal in the form

$$Z_S(t) = AZ_C(t)\exp(j\Omega t), \tag{53.6}$$

where A is a constant characterizing the level of the echo signal. And, finally, we designate the energy of the emitted signal by

$$E_C = \int_{-\infty}^{\infty} |Z_C(t)|^2\, dt. \tag{53.7}$$

We define the mathematical expectation $m(1)$, which on the basis of (52.5), and (53.4)-(53.7) equals

$$m(1) = AE_C. \tag{53.8}$$

The variance

$$d(1) \equiv \langle |Y(1)|^2\rangle - |\langle Y(1)\rangle|^2 \tag{53.9}$$

is found by substituting (53.4) into (53.9). Taking account of (53.5)-(53.7) we see that substitution leads to

$$d(1) = \int_{-\infty}^{\infty}\int_{-\infty}^{\infty} B_F(t'-t'')\, Z_C^*(t')\, Z_C(t'')\exp[j\Omega(t'-t'')]\, dt'\, dt'' + \\ + \int_{-\infty}^{\infty}\int_{-\infty}^{\infty} B_N(t'-t'')\, Z_C^*(t')\, Z_C(t'')\exp[j\Omega(t'-t'')]\, dt'\, dt'', \tag{53.10}$$

where

$$B_F(t'-t'') = \langle Z_F(t')\, Z_F^*(t'')\rangle = I_F R_F(t'-t''); \tag{53.11}$$

$$B_N(t'-t'') = \langle Z_N(t')\, Z_N^*(t'')\rangle \tag{53.12}$$

are autocorrelation functions for reverberation and noise interference, respectively.

We note that (53.11) is valid when reverberation is a random process reducible to a wide-sense stationary state (see Chapter VI) -- that is, when its normalized autocorrelation function $R_F(t' - t'')$ depends only on the time difference.

In accordance with §40 we now have

$$B_F(t'-t'')=\frac{I_F}{E_C}\int_{-\infty}^{\infty} Z_C(t)Z_C^*(t+t'-t'')\,dt, \tag{53.13}$$

where I_F is the average intensity of the reverberation.

On the basis of (53.10)-(53.13) we next obtain the following relationships for the variances $d(1)$ and $d(0)$:*

$$d(1)=I_F E_C\int_{-\infty}^{\infty}|\chi(\Omega,\tau)|^2\,d\tau+G_N E_C; \tag{53.14}$$

$$d(0)=I_F E_C\int_{-\infty}^{\infty}|\chi(\Omega,\tau)|^2\,d\tau+G_N E_C, \tag{53.15}$$

where

$$\chi(\Omega,\tau)=\frac{1}{E_C}\int_{-\infty}^{\infty} Z_C(t)Z_C^*(t+\tau)\exp(j\Omega t)\,dt \tag{53.16}$$

is the ambiguity function of the emitted signal (see §22).

Considering that $d = d(1) = d(0)$, we see that the detection characteristics (49.9) is specified by

$$P_{\mathrm{d}}=\frac{1}{2}\left\{1+\Phi\left[\frac{m(1)}{\sqrt{d}}-\Phi^{-1}(1-2P_{\mathrm{fa}})\right]\right\}. \tag{53.17}$$

The relationship (53.17) contains the parameter $m(1)/\sqrt{d}$, which according to (53.8), (53.14), and (53.15) is equal to

$$\frac{m(1)}{\sqrt{d}}=\frac{A\sqrt{E_C}}{\sqrt{I_F\int_{-\infty}^{\infty}|\chi(\Omega,\tau)|^2\,d\tau+G_N}}. \tag{53.18}$$

As an example, we will examine a bell-shaped emitted signal with linear frequency modulation. Considering the

---

*See (50.3) for the definition of G_N (= G) here. (D.M.)

data in § 42, we write

$$\int_{-\infty}^{\infty} |\chi(\Omega, \tau)|^2 d\tau = \frac{\sqrt{2}\, T_{ef}}{\mu} \exp\left(-\frac{2\pi\nu^2}{\mu^2}\right), \tag{53.19}$$

where

$$T_{ef} = E_C \tag{53.20}$$

is the effective signal duration (its amplitude is normalized and equal to one);

$$\mu = \Delta F_{ef} T_{ef} = \sqrt{1 + 4(\Delta F_m T_{ef})^2} \tag{53.21}$$

is the signal complexity factor; ΔF_{ef} is the effective width of the spectrum; ΔF_m is the effective depth of frequency modulation; and

$$\nu = \frac{\Omega T_{ef}}{2\pi} \tag{53.22}$$

is the relative frequency shift produced by the Doppler frequency shift connected with movement of the object of detection.

Now, substituting (53.19) into (53.18), we get

$$\frac{m(1)}{\sqrt{d}} = \frac{A}{\sqrt{\frac{\sqrt{2}\, I_F}{\mu} \exp\left(-\frac{2\pi\nu^2}{\mu^2}\right) + \frac{G_N}{T_{ef}}}}. \tag{53.23}$$

We note that A is interpreted as the amplitude of the echo signal, and we introduce the parameters q_{SN} and q_{FN} respectively defining the average echo signal intensity to noise interference intensity ratio and the reverberation interference to noise interference intensity ratio:

$$q_{SN} = \frac{I_S}{I_N} = \frac{A^2 T_{ef}}{2G_N}; \tag{53.24}$$

$$q_{FN} = \frac{I_F}{I_N} = \frac{I_F T_{ef}}{G_N}. \tag{53.25}$$

Then it follows from (53.23)-(53.25) that

$$\frac{m(1)}{\sqrt{d}} = \frac{\sqrt{2q_{SN}}}{\sqrt{q_{FN}\frac{\sqrt{2}}{\mu}\exp\left(-\frac{2\pi\nu^2}{\mu^2}\right)+1}}. \tag{53.26}$$

Relationship (53.17) shows that the parameter $m(1)/\sqrt{d}$ uniquely defines the echo signal detection characteristic at a fixed value of P_{fa}. Therefore, it is of interest to examine the dependence of this parameter on the other parameters in formula (53.26). We are most interested in evaluating the effects of two parameters -- the relative Doppler frequency shift ν and the emitted signal complexity factor μ. We note that in view of (53.21) the value of μ can change as a result of changes in either ΔF_{ef} or T_{ef}. We will assume that μ changes in response to a change in ΔF_{ef}, since T_{ef} would also affect q_{SN} and q_{FN} under certain conditions, which would complicate this examination.

Figure 15 shows the dependence of $m(1)/\sqrt{d}$ on the parameter μ at different values of ν. In this case, q_{SN} and

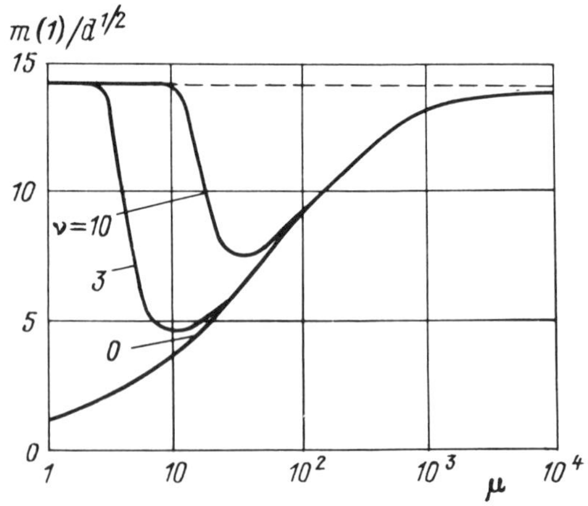

Figure 15. Effect of the type of emitted signal on the signal/interference ratio, Eq. (53.26), at the output of a cross-correlation detection system, with q_{SN} = 20 db, q_{FN} = 20 db.

q_{FN} are assumed constant. It follows from these plots that:

(a) at $\nu = 0$ -- that is, in the absence of Doppler frequency shift in the echo signal, echo signal detection improves as the complexity factor increases, owing to a reduction of the effect of reverberation interference;

(b) at $\nu \neq 0$ there are such values of μ at which echo signal detection is poorest, owing to the heightened effect of reverberation interference;

(c) the greater the size of ν, the more values of μ there are at which the echo signal detection conditions are poorest.

§54. Detection with Multichannel Systems, and Discussion of Results

Up to now in discussing echo signal detection characteristics in the presence of interference we have been talking about so-called "point detection." This is defined as making a decision as to the presence or absence of an echo signal on the basis of observations on random variable Y. This random variable corresponded to a definite moment at which an echo signal may appear (range to the object of detection), one frequency shift (rate of movement of the object of detection), and one directional pattern (direction to the object of detection). But at the same time a sonar system operates in a space-frequency-time domain and, consequently, the value of the adopted false alarm probability P_{fa} must take this circumstance into account. The question arises: What value should we select for the probability of false alarms in analyzing echo signal detection characteristics? We will assess this value.

Let the sonar system have N_s spatial channels, with each containing N_f frequency channels. It follows from this that the total number of observation channels N in the sonar system equals

$$N = N_s N_f. \tag{54.1}$$

It is not difficult to show that at $P_{fa} \ll 1$ and when the false alarms in different channels are independent, the probability of false alarms $P_{fa}^{(N)}$ in an N-channel system is associated with the probability of false alarms P_{fa} "at a point" by the relationship

$$P_{fa}^{(N)} = NP_{fa}, \tag{54.2}$$

or, taking account of (54.1), by

$$P_{fa}^{(N)} = N_s N_f P_{fa}.$$

In this case we assume that the probability of false alarms is the same for all channels. If this is not so and if the probability of false alarms equals P_{fak} for any kth channel, then instead of (54.2) the following general relationship holds:

$$P_{fa}^{(N)} = 1 - \prod_{k=1}^{N} (1 - P_{fak})$$

or, on the condition that $P_{fak} \ll 1$,

$$P_{fa}^{(N)} = \sum_{k=1}^{N} P_{fak}. \tag{54.3}$$

Next, let us examine the temporal dynamics of the appearance of false alarms. Assuming that false alarms follow Poisson's law, we have

$$P(k) = \frac{(n_1 T)^k}{k!} \exp(-n_1 T), \tag{54.4}$$

for the probability $P(k)$ that k false alarms would appear in one channel during time T, where n_1 is the average number of false alarms per unit time. If we assume that $k = 0$, then

$$P(0) = \exp(-n_1 T_0), \tag{54.5}$$

in which case T_0 can be interpreted as the average time during which the probability is $P(0)$ that not a single false alarm arises.

The value of n_1 can be defined for different probability distributions of the process $Y(t)$ at the output of one channel in the sonar system. In particular, if the process $Y(t)$ follows a Rayleigh probability distribution, then we have the known relationship for n_1 [12]

$$n_1 = \frac{2\sqrt{\pi}\, P_{\mathbf{fa}} \sqrt{\ln \frac{1}{P_{\mathbf{fa}}}}}{\tau_k}, \tag{54.6}$$

where τ_k is the correlation interval for interference in $Y(t)$. Substituting (54.6) into (54.5) we get

$$P(0) = \exp\left(-\frac{T_0}{\tau_k} 2\sqrt{\pi}\, P_{\mathbf{fa}} \sqrt{\ln \frac{1}{P_{\mathbf{fa}}}}\right). \tag{54.7}$$

We are actually interested in false alarms in an N-channel system. Therefore in the general case, when $N > 1$, formula (54.7) should be rewritten as

$$P(0) = \exp\left(-\frac{T_0}{\tau_k} 2\sqrt{\pi}\, P_{\mathbf{fa}}^{(N)} \sqrt{\ln \frac{1}{P_{\mathbf{fa}}^{(N)}}}\right), \tag{54.8}$$

or, considering (54.3), as

$$P(0) = \exp\left(-\frac{T_0}{\tau_k} 2\sqrt{\pi}\, N P_{\mathbf{fa}} \sqrt{\ln \frac{1}{N P_{\mathbf{fa}}}}\right). \tag{54.9}$$

Relationship (54.9) permits us to determine the permissible value of the probability of false alarms P_{fa} "at a point," as had been done in examining the detection characteristics in the preceding paragraphs of this chapter. This determination can be made by using easily interpretable initial data. The following expression is somewhat more convenient than (54.9):

$$\frac{\tau_k}{T_0} = \frac{2\sqrt{\pi}\, N P_{\mathbf{fa}}}{\ln \frac{1}{P(0)}} \sqrt{\ln \frac{1}{N P_{\mathbf{fa}}}}. \tag{54.10}$$

Unfortunately, equation (54.10) is a transcendental relation in the quantity of interest to us, P_{fa}, and it must be solved either graphically or by successive selection of P_{fa} values which agree with the values assumed for the rest of the parameters. We can, however, point out an approximate expression for P_{fa} which is valid for $NP_{\mathrm{fa}} = 10^{-2}$–$10^{-8}$:

$$P_{\mathrm{fa}} \approx \frac{\tau_k}{10NT_0} \ln \frac{1}{P(0)}. \tag{54.11}$$

Moreover, if probability $P(0)$ is close to one, then we can assume as an approximation that

$$\ln \frac{1}{P(0)} \approx 1 - P(0). \tag{54.12}$$

Then, we get the final expression for the probability of false alarms "at a point" from (54.11) and (54.12):

$$P_{\mathrm{fa}} \approx \frac{\tau_k}{10NT_0} [1 - P(0)]. \tag{54.13}$$

As an example, we examine the case in which $P(0) = 0.9$, $T_0 = 10$ min, $\tau_k = 1$ sec, and $N = 100$. In this case we find that the required probability of false alarms is $P_{\mathrm{fa}} \simeq 2 \cdot 10^{-7}$.

And so, prescribing easily interpretable parameters -- probability $P(0)$ that not one false alarm occurs during a time interval τ_k, the correlation interval T_0 of interference, and the number of independent channels N in the sonar system -- we can find the needed probability of false alarms "at a point" P_{fa}. After this, we use one of the graphs in Figures 11-14 which corresponds to the sonar information processing system being used and to the adopted probability models for the signals and interference, and we determine the signal/interference ratio q at which the needed detection probabilities P_O are maintained. Naturally, the adequacy with which the adopted model of the input process reflects real sonar conditions must be checked experimentally using quantitative methods of research (see Chapter II).

In conclusion, we note that in using parameters q_1, q_2, and q_3 we must take account of differences in their definitions, as given by formulas (50.2), (50.3), (50.5), and (51.22), and in the footnote on p. 162.

CHAPTER VIII

ELEMENTS OF THE THEORY OF STATISTICAL MEASUREMENT

§55. Basic Stages of Statistical Measurements

As had been noted in §5 and §6, experimental research can be either qualitative or quantitative. A necessary element of quantitative experimental research is the performance of statistical measurement, which is defined as obtaining statistical estimates of the probability characteristics (statistics, for short) of random processes, with a known degree of accuracy. We will examine below some aspects of statistical measurement, as defined here. We will use designations and terms adopted in Chapter II. The material of this chapter is based on [3,8,9,18,21,23,24].

Statistical measurement should be performed with the assistance of a statistical measuring system containing the following basic elements -- a device for memorizing and storing the selective process $\hat{x}(t)$, a selective process classifier, an estimator $\hat{\Theta}(l)$ of the probability characteristic $\Theta(l)$, a statistical measurement optimizer, and a device for representing the results of statistical measurement (cf. Figure 2).

Let us examine the purpose and function of these elements in greater detail.

Obtained by experiment, the selective process $\hat{x}(t)$ is fed into the memory. This process can be initially sampled in time or quantized in level. When necessary, $\hat{x}(t)$ can be expanded on the basis of a relevant system of functions. Its envelope and phase, its quadrature components, and so on, can be isolated. In a number of cases, when storing the selective process we also encounter the problem of minimizing the description of input data, since the volume of such data may be extremely large, making their subsequent use difficult.

The following points should be considered when we subject the selective process to transformation prior to its storage in the memory. Let the function

$$\hat{x}_R(t) = R[\hat{x}(t)] \tag{55.1}$$

be the result of transformations of the function $\hat{x}(t)$ by the operator R. If W is an operator describing retrieval of the initial process $\hat{x}(t)$ from data on $\hat{x}_R(t)$, then we can introduce the *retrieval error*

$$\rho_{RW} = \rho[\hat{x}(t), \hat{x}_R(t)], \tag{55.2}$$

which characterizes the difference between the processes W with respect to criterion ρ. The value of this error should agree with the resultant error in statistical measurement, and it should be selected on the basis of the overall task of measurement (this point will be discussed in subsequent paragraphs of this chapter). Thus, the type of transformation operator R must correspond to the purpose of statistical measurement, and its selection is not arbitrary.

After the process $\hat{x}(t)$ is fed into the memory, it must be classified. At least (see §14), we must indicate to which of the following basic classes of random processes the selective process $\hat{x}(t)$ belongs -- nonstationary heterogeneous processes, stationary heterogeneous, nonstationary homogeneous, or stationary homogeneous (ergodic). Such a classification enables us to select the type of probability characteristic $\Theta(l)$ to be studied and, in addition it affects the selection of algorithms by which to form the statistical estimates $\hat{\Theta}(l)$. We determine the type of probability characteristic to be studied on the basis of the class of the process being studied and on the specific ways we intend using the results of statistical measurement, as prescribed by the experimental goals. It may be current-t $\Theta(t,l)$, current-k $\Theta(k,l)$, or average $\Theta(l)$. We note that, for the moment, we are discussing not the form of the characteristic but rather its type, since the form is defined by the class only partially.

The next stage in statistical measurement is computing the statistical estimate $\Theta(l)$. Prior to the start of a statistical measurement the experimenter does not know the specific values of the probability characteristic $\Theta(l)$ under investigation. Therefore, he does not have adequate grounds on which to select the parameters of the measuring system (valuation estimator). However, after the estimate $\hat{\Theta}(l)$ is obtained, the possible form of $\Theta(l)$ is defined much more precisely. This naturally brings up the question of using adaptive methods for optimizing statistical measurement. Such a procedure should be taken to mean subjective selection of the system's parameters, acquisition of initial statistical estimates, and optimization of the measurements, adaptively. It is precisely the statistical measurement optimizer that carries out this procedure.

Checking the statistical measuring system with the assistance of standard random processes is an important stage of statistical measurement. After we classify the selective process $\hat{x}(t)$ and optimize the measurements, we can select a standard process which corresponds to the needed class of selective process $\hat{x}(t)$ and the form of the estimate $\hat{\Theta}(l)$. This enables us to determine the magnitude of the measurement errors. Consequently, this means that we can justify the work of the statistical measuring system from a methodological standpoint. The collection of standard processes must be rather extensive, and it should satisfy the following requirement: For processes of a particular class, the collection of standard processes must encompass all typical forms of characteristics $\Theta(l)$.

The last stage of statistical measurement is representing the estimate $\hat{\Theta}(l)$. In this case, we can represent the estimate directly, or we can approximate it by a corresponding system of functions. The form in which we compare the estimate $\hat{\Theta}(l)$ with the probability characteristic $\Theta(l|m)$, corresponding to model m of the random process, is important in this regard. Since the final goal of experiments is to select the base model from among the possible models $m \in M$ (see §8), the form in which the estimate $\hat{\Theta}(l)$ is represented must match the form of $\Theta(l|m)$. These forms can include graphs, sets of digital values, or analytical expressions.

Thus, measurement of the probability characteristic of a random process under investigation contains the following basic stages:

1. identification of the selective process with one of the classes;

2. computation of the statistical estimate for the characteristic under investigation;

3. optimization of the statistical measurement procedure;

4. acquisition of quantitative values for the measured characteristic with controllable accuracy;

5. comparison of the statistical estimate with the possible probability characteristics, and determination of the base model for the random process.

§56. Statistical Measurement Errors*

Classifying statistical measurement errors is most important from a methodological standpoint. As a matter of course, it is precisely the classification of errors and investigation of their basic properties that permit statistical measurement to be a method for quantitatively determining the probability characteristics of random processes.

Statistical measurement errors can be classified on the basis of two traits -- the nature of their origin, and their manifestations.

Statistical measurement errors are divided into four types with respect to the nature of their origin:

1. errors in representing the selective process;

2. errors in classifying the selective process;

3. algorithmic errors;

4. errors inherent to the measuring apparatus.

Errors of the first type are connected with the representation of the selective process, and they arise when the process is transformed prior to its introduction into

---

*[53] and references therein; see also [68]. (D.M.)

the statistical measuring system. Some aspects of this problem have been discussed in §55.

Errors of the second type are connected with the lack of correspondence between the selective process and the model adopted for the random process under investigation. Errors of this type have been treated in Chapter II, where we have discussed the problems of setting up and interpreting experimental research.

Errors of the third type arise as a consequence of incorrect selection of the algorithm for forming the statistical estimate as compared to the algorithm for the determination of the probability characteristic under investigation -- that is, as compared to the mathematical definition of this characteristic. Algorithmic errors have their source primarily in the parameters of the statistical error computer. Errors of this type will be discussed in subsequent paragraphs of this chapter.

Errors of the fourth type arise as a result of imprecise performance of the adopted algorithms for computing statistical estimates by the particular statistical measuring equipment.

As has been noted, statistical measurement errors can also be classified with respect to their manifestations or their specific nature. In this case, we distinguish between fluctuating errors and errors involving bias in the statistical estimates.

Fluctuating errors have a random nature. Their values change as we go from one measuring cycle to another. Errors stemming from bias in statistical estimates or, as they are called for short, estimation bias, have a systematic nature, and they are repeated in different measuring cycles. Of course, this does not mean that estimation bias is necessarily known and that it can be compensated for in some way, to reduce it to zero.

In principle, all four types of error mentioned above (representation, classification, algorithmic, and apparatus errors) should contribute to both fluctuating error and estimation bias. Examination of these errors and determining ways to minimize them is the key problem of the theory of statistical measurement.

§57. Formation of a Statistical Estimate

As follows from our examination of the structure of the statistical measuring system (see §55), we form a statistical estimate $\hat{\Theta}(l)$ of the probability characteristic $\Theta(l)$ with the assistance of an estimator. If $x(t)$ is a selective process obtained by experiment and $\{\hat{x}_k(t)\}$ is a selective representation of this process, then in the general case the statistical estimate $\hat{\Theta}(l)$ is formed as follows.

Let ϑ be an operator for transforming the random process of interest, $x(t)$, corresponding to the selected probability characteristic Θ. Limiting ourselves to the case of one-dimensional probability characteristics, such that $l = (k,t,l)$, we define the estimate $\hat{\Theta}(k,t,l)$ in the following way:

$$\hat{\Theta}(k, t, l) = \sum_{k'=-\infty}^{\infty} h(k, k') \int_{-\infty}^{\infty} \int_{-\infty}^{\infty} h(t, t') h(l, l') \vartheta [\hat{x}_k(t'), l'] \, dl' \, dt', \quad (57.1)$$

where $h(k,k')$, $h(t,t')$, and $h(l,l')$ are weight functions corresponding to the number k' of the selective processes, current time t', and variable l', which govern the definition of the one-dimensional characteristic of interest to us. We introduce the functions h into the expression (57.1) to emphasize the possibility for averaging when we compute an estimate having unequal weight factors with respect to the arguments k, t, and l. At the same time, we take account of the possibility for different domains of these arguments in forming the estimate. We note that the selection of the form of the function h is an interesting optimization problem of statistical measurement. However, we will limit ourselves here to the problem of parametric optimization, and we will examine the so-called "centiform" weight functions, which have the following form:

$$\left.\begin{aligned} h(k, k') &= \frac{1}{2n+1}, \quad k-n \leqslant k' \leqslant k+n; \\ h(t, t') &= \frac{1}{T}, \quad t-\frac{T}{2} \leqslant t' \leqslant t+\frac{T}{2}; \\ h(l, l') &= \frac{1}{L}, \quad l-\frac{L}{2} \leqslant l' \leqslant l+\frac{L}{2}, \end{aligned}\right\} \quad (57.2)$$

where $2n + 1 = N$ is the number of selective representations being averaged, T is the temporal smoothing interval, and L is the resolution with respect to the argument l. Substituting (57.2) into (57.1), we get the following particular representation of the estimate $\hat{\Theta}(l)$:

$$\hat{\Theta}(k, t, l) = \frac{1}{NTL} \sum_{k'=k-n}^{k+n} \int_{t-T/2}^{t+T/2} \int_{l-L/2}^{l+L/2} \vartheta[\hat{x}_{k'}(t'), l']\, dl'dt'. \qquad (57.3)$$

The relation (57.3) demonstrates that the estimate $\hat{\Theta}(k,t,l)$ is formed by "symmetrical" averaging with respect to $N = 2n + 1$ selective representations, for which $\hat{x}_k(t)$ is the average, with respect to time over the interval relative to the mean time $t = t'$, and with respect to argument l' over the interval L, which has $l = l'$ as its mean.

A particular feature of the estimate $\hat{\Theta}(k,t,l)$ is that in the general case it is a random function of its arguments -- representation number k, reading time t, and argument l. This is natural, since the result of a measuring procedure in statistical measurement is always random, in view of the incomplete knowledge of the model of the process being studied and the impossibility of predicting its values precisely.

Some remarks on the resolution L are in order. In this case we limit ourselves to an examination of the current-t (i.e., instantaneous) probability characteristics, such as the probability density $W(t,X)$, the correlation function $B(t,\tau)$, and energy spectrum $G(t,\omega)$. Here the resolution L is assumed to be the differential interval $\Delta x = L$ when measuring $W(t,X)$, the interval of correlation shifts $\Delta\tau = L$ when measuring $B(t,\tau)$, and the frequency interval $\Delta\omega = L$ when measuring $G(t,\omega)$. As we will see below, the reason we introduce smoothing with respect to the arguments X, τ, or ω is to decrease the statistical measurement error.

§58. Statistical Measurement Errors of the Characteristics of Nonstationary Homogeneous Random Processes

A nonstationary homogeneous random process is distinguished by the fact that its current-k probability characteristic does not depend on the number k of the selected representation and, consequently,

$$\Theta(k, t, l) = \Theta(t, l). \tag{58.1}$$

In this case the current-t characteristic

$$\Theta(t, l) = \langle \vartheta [x_k(t), l] \rangle \tag{58.2}$$

bears the principal information regarding the nonstationary properties of the process under investigation. Marine reverberation and echo signals are typical processes of this type in sonar [2,15-17,19,33].

Let us define the measurement error $\rho_\Theta(t,l)$ in the following form:

$$\rho_\Theta(t, l) = \langle [\hat{\Theta}(t, l) - \Theta(t, l)]^2 \rangle, \tag{58.3}$$

where $\hat{\Theta}(t,l)$ is the statistical estimate of the current-probability characteristic $\Theta(t,l)$. The value of $\rho_\Theta(t,l)$ corresponds to the mean square of the total statistical measurement errors and includes both the fluctuating error and the estimate bias [18,23,24]. Next, we substitute expression (57.3) into (58.3):

$$\rho_\Theta(t, l) = \left\langle \left\{ \frac{1}{NTL} \sum_{k'=k-n}^{k+n} \int_{t-T/2}^{t+T/2} \int_{l-L/2}^{l+L/2} \vartheta [\hat{x}_{k'}(t'), l'] \, dl' dt' - \Theta(t, l) \right\}^2 \right\rangle. \tag{58.4}$$

Squaring the expression contained within braces in (58.4) and averaging term by term, we get

$$\rho_\Theta(t, l) = \frac{1}{(NTL)^2} \sum_{k'=k-n}^{k+n} \sum_{k''=k-n}^{k+n} \iint_{t-T/2}^{t+T/2} \iint_{l-L/2}^{l+L/2} \langle \vartheta [\hat{x}_{k'}(t') l'] \vartheta [\hat{x}_{k''}(t''), l'] \rangle \times$$
$$\times dl' dl'' dt' dt'' - \frac{2\Theta(t, l)}{NTL} \sum_{k'=k-n}^{k+n} \int_{t-T/2}^{t+T/2} \int_{l-L/2}^{l+L/2} \langle \vartheta [\hat{x}_{k'}(t'), l'] \rangle \, dl' dt' + \Theta^2(t, l). \tag{58.5}$$

Let us examine the function

$$I(k', k''; t', t''; l', l'') \equiv \langle \vartheta [\hat{x}_{k'}(t'), l'] \vartheta [\hat{x}_{k''}(t''), l''] \rangle, \tag{58.6}$$

located after the summation and integral signs of the first term in expression (58.5). Assuming that the selective representations $\hat{x}_k(t)$ are statistically independent at

$k' \neq k''$, we have

$$I(k', k''; t', t''; l', l'') = \begin{cases} \langle \vartheta [\hat{x}_k(t'), l'] \vartheta [\hat{x}_k(t''), l''] \rangle, & k' = k'' = k; \\ \langle \vartheta [\hat{x}_{k'}(t'), l'] \rangle \langle \vartheta [\hat{x}_{k''}(t''), l''] \rangle, & k' \neq k''. \end{cases} \tag{58.7}$$

This assumption is justified in practice since under most typical conditions of experimental research, different representation numbers, k, correspond to different experiments, the conditions of which are independent due either to the way the experiments are set up or to the application of known methods of randomization. In this case, we define randomization as the creation of independent conditions under which experiments are conducted. In sonar, in particular, randomization can be achieved through proper selection of intervals between successive emission-reception cycles, variation of the central frequency of the emitted signals, selection of the working frequency ranges, and so on.

We will assume that errors in transforming the selective representations $\hat{x}_k(t)$ are absent and that the model of the process $x(t)$ has been developed correctly. This means that classification errors are absent, as well, Considering that we are examining homogeneous random processes, we note that the conditions we have formulated enable us to write the equality

$$\Theta(t, l) = \langle \vartheta [\hat{x}_k(t), l] \rangle \tag{58.8}$$

for any number k of selective representations $\hat{x}_k(t)$. The relationship (58.8) indicates that the model of the process under investigation does not depend on the number, k, of selective representations $\hat{x}_k(t)$. Of course, this imposes specific requirements on the way the experiments are set up. The conditions under which they are conducted must be statistically homogeneous, and they should ensure independence of the model of the phenomenon from the number k of the experiment [53].

Considering (58.7) and (58.8), we have

$$I(k', k''; t', t''; l', l'') = \begin{cases} B_\vartheta(t', t''; l', l'') + \Theta(t', l')\Theta(t'', l''), & k' = k'' = k; \\ \Theta(t', l')\Theta(t'', l''), & k' \neq k'', \end{cases} \tag{58.9}$$

where

$$B_\vartheta(t', t''; l', l'') = \langle\{\vartheta[\widehat{x_k}(t'), l'] - \Theta(t', l')\}\{\vartheta[\widehat{x_k}(t''), l''] - \Theta(t'', l'')\}\rangle \quad (58.10)$$

is the correlation function for fluctuations of the operator $\vartheta[\hat{x}_k(t), l]$. The correlation function (58.10) can be interpreted as follows: It establishes the correlative associations of the process $\vartheta[\hat{x}_k(t), l]$, which is obtained by the transformation of the representations $\hat{x}_k(t)$ by the operator ϑ for different values of t and l. Under the conditions assumed, in which transformation and classification errors are absent, the following equality holds:

$$\vartheta[\widehat{x_k}(t), l] = \vartheta[x_k(t), l].$$

On the basis of relationships (58.5)-(58.10) we get the following expression for the square of the total statistical measurement error:

$$\rho_\Theta(t, l) = \frac{1}{N(TL)^2}\int\limits_{t-T/2}^{t+T/2}\int\limits_{t-T/2}^{t+T/2}\int\limits_{l-L/2}^{l+L/2}\int\limits_{l-L/2}^{l+L/2} B_\vartheta(t', t''; l', l'')\,dl'\,dl''\,dt'\,dt'' + \left[\frac{1}{TL}\int\limits_{t-T/2}^{t+T/2}\int\limits_{l-L/2}^{l+L/2}\Theta(t', l')\,dl'\,dt' - \Theta(t, l)\right]^2. \quad (58.11)$$

The first term of expression (58.11) is nothing more than the variance of the fluctuation error of the statistical measurement

$$d_\Theta(t, l) = \frac{1}{N(TL)^2}\int\limits_{t-T/2}^{t+T/2}\int\limits_{t-T/2}^{t+T/2}\int\limits_{l-L/2}^{l+L/2}\int\limits_{l-L/2}^{l+L/2} B_\vartheta(t', t''; l', l'')\,dl'\,dl''\,dt'\,dt'', \quad (58.12)$$

while the second term is the square of the estimate bias

$$S_\Theta(t, l) = \frac{1}{TL}\int\limits_{t-T/2}^{t+T/2}\int\limits_{l-L/2}^{l+L/2}\Theta(t', l')\,dl'\,dt' - \Theta(t, l). \quad (58.13)$$

Thus, the total mean-square error [based on the definition (58.3)], e.g.,

$$\rho_\Theta(t, l) = d_\Theta(t, l) + S_\Theta^2(t, l), \quad (58.14)$$

consists of the variance of the fluctuation error and the square of the estimate bias.

Let us return to expression (58.12) for the fluctuation error. We represent the correlation function $B_\vartheta(t',t''; l',l'')$ in the following form:

$$B_\vartheta(t',t'';l',l'') = d_\vartheta^{1/2}(t)\, d_\vartheta^{1/2}(l)\, R_t(t'-t'')\, R_l(l'-l''), \qquad (58.15)$$

where $d_\vartheta(t)$ and $d_\vartheta(l)$ are the variances of the fluctuations in $\vartheta[\hat{x}_k(t), l]$ with respect to coordinates t and l respectively, and where $R_t(t' - t'')$ and $R_l(l' - l'')$ are the normalized autocorrelation functions of this operator. The expression (58.15) is valid in the event that the temporal nonstationariness and variation with respect to argument are adequately "slow." Such an assumption on the "slowness" of fluctuations requires justification for each specific case of research. This problem can be solved by using the probability model of the process under investigation, in which case we usually can prove the validity of the following representation:

$$x(t) = \xi(t)\, f(t),$$

where $\xi(t)$ is a stationary homogeneous random process and $f(t)$ is a slowly changing function. We note that, strictly speaking, we must prove the "slowness" of the transformed $\vartheta[\xi(t), l]$ and $\vartheta[f(t), l]$. This condition is satisfied in sonar for marine reverberation, echo signals, and some types of underwater noise [2,14,17,22].

Substituting (58.15) into (58.12) and reducing the double integrals to single integrals produces

$$d_\Theta(t, l) = \frac{4d_\vartheta^{1/2}(t)\, d_\vartheta^{1/2}(l)}{NTL} \int_0^T \left(1 - \frac{\tau}{T}\right) R_t(\tau)\, d\tau \int_0^L \left(1 - \frac{\lambda}{L}\right) R_l(\lambda)\, d\lambda. \qquad (58.16)$$

Next, let us examine a case in which the following inequalities hold:

$$\tau_\vartheta = \int_0^T R_t(\tau)\, d\tau \gg \int_0^T \frac{\tau}{T} R_t(\tau)\, d\tau; \qquad (58.17)$$

$$\lambda_\vartheta = \int_0^L R_l(\lambda)\,d\lambda \gg \int_0^L \frac{\lambda}{T} R_l(\lambda)\,d\lambda, \tag{58.18}$$

where τ_ϑ and λ_ϑ are the correlation intervals for the fluctuations in the operator ϑ with respect to coordinates t and l, respectively. Conditions (58.17) and (58.18) correspond to cases in which the correlation intervals τ_ϑ and λ_ϑ are much smaller than the smoothing time T and the resolution L used in making the statistical measurements. These conditions usually hold.

On the basis of (58.16)-(58.18), we get the following final expression for the variance $d_\Theta(t, l)$:

$$d_\Theta(t, l) = \frac{4d_\vartheta^{1/2}(t)\,d_\vartheta^{1/2}(l)\,\tau_\vartheta l_\vartheta}{NTL}. \tag{58.19}$$

This expression can be rewritten in somewhat different form upon assuming

$$n_T = \frac{T}{\tau_\vartheta}, \quad n_L = \frac{L}{l_\vartheta} \tag{58.20}$$

to be the number of uncorrelated readings within the limits of the smoothing time T and the resolution L, correspondingly. Then, obviously,

$$d_\Theta(t, l) = \frac{4d_\vartheta^{1/2}(t)\,d_\vartheta^{1/2}(l)}{Nn_T n_L}. \tag{58.21}$$

It follows from (58.21) in particular that the variance of the fluctuation error decreases as the number of averaged representations N and the number of uncorrelated readings n_T and n_L increase, within the limits of averaging time T and resolution L.

Next, let us examine the $S_\Theta^2(t, l)$ component of the total measurement error $\rho_\Theta(t, l)$. We make the following substitutions, $\tau = t' - t$, $\lambda = l' - l$ in the integrand of (58.13) in order to obtain a simple expression for the estimate bias $S_\Theta(t, l)$. Then,

$$S_\Theta(t, l) = \frac{1}{TL} \int_{-T/2}^{T/2} \int_{-L/2}^{L/2} \Theta(t + \tau, l + \lambda)\,d\lambda\,d\tau - \Theta(t, l). \tag{58.22}$$

We expand the function $\Theta(t+\tau, l+\lambda)$ into a double Taylor series in the neighborhood of points t and l in terms of powers of τ and λ:

$$\Theta(t+\tau, l+\lambda) = \sum_{p=0}^{\infty}\sum_{q=0}^{\infty} \frac{\Theta^{(p,q)}(t,l)}{p!\,q!}\tau^p\lambda^q, \qquad (58.23)$$

where

$$\Theta^{(p,q)}(t,l) = \frac{\partial^p\partial^q}{dt^p dl^q}\Theta(t,l) \qquad (58.24)$$

are derivates of orders p and q with respect to the variables t and l, respectively. Now, substituting (58.23) into (58.22), taking account of the fact that $\Theta^{(0,0)}(t,l) = \Theta(t,l)$, and integrating, we get

$$S_\Theta(t,l) = 8\sum_{p=1}^{\infty}\frac{\Theta^{(2p,0)}(t,l)}{(2p+1)!}\left(\frac{T}{2}\right)^{2p} + 8\sum_{q=1}^{\infty}\frac{\Theta^{(0,2q)}(t,l)}{(2q+1)!}\left(\frac{L}{2}\right)^{2q} + 8\sum_{p=1}^{\infty}\sum_{q=1}^{\infty}\frac{\Theta^{(2p,2q)}(t,l)}{(?p+1)!\,(2q+1)!}\left(\frac{T}{2}\right)^{2p}\left(\frac{L}{2}\right)^{2q}. \qquad (58.25)$$

If we consider the first two terms in the expansion (58.25), then for $S_\Theta(t,l)$ we have

$$S_\Theta(t,l) = \frac{\Theta^{(2,0)}(t,l)}{24}T^2 + \frac{\Theta^{(0,2)}(t,l)}{24}L^2. \qquad (58.26)$$

As can be seen from (58.26), the estimate bias $S_\Theta(t,l)$ depends on the value of the second derivatives $\Theta^{(2,0)}(t,l)$ and $\Theta^{(0,2)}(t,l)$, and it increases as the averaging time T and the resolution increases. Substituting expressions (58.19) and (58.26) into (58.14), we get the following for the mean square $\rho_\Theta(t,l)$ of the total error:

$$\rho_\Theta(t,l) = \frac{4d_\vartheta^{1/2}(t)\,d_\vartheta^{1/2}(l)\,\tau_\vartheta l_\vartheta}{NTL} + \left[\frac{\Theta^{(2,0)}(t,l)}{24}T^2 + \frac{\Theta^{(0,2)}(t,l)}{24}L^2\right]^2. \qquad (58.27)$$

The relationship (58.27) demonstrates that different components of the statistical measurement error depend differently on parameters T and L of the measuring system, as well as on the number of averaged representations N. In fact, as the number of averaged representations N increases, the total mean-square error $\rho_\Theta(t,l)$ decreases, asymptotically approximating $S^2(t,l)$, which does not depend on N. In this case, the total error decreases in response to a decrease in the variance of the fluctuation error $d_\Theta(t,l)$, the value of which is inversely proportional to N. When the smoothing time T or parameter L, which characterizes resolution with respect to argument l, is increased, the total error reaches a minimum. This occurs because as parameters T and L are increased, the variance of the fluctuation error drops at the same time that the estimate bias rises. It is important to mention the conditions under which these dependencies of bias $S_\Theta(t,l)$ on T and L hold. If

$$\Theta^{(2,0)}(t, l)/\Theta^{(0,2)}(t, l) > 0, \tag{58.28}$$

that is, if the second derivatives of the probability characteristic $\Theta(t,l)$ have the same sign, then the nature of the dependence of the bias on T and L is as noted above. But if

$$\Theta^{(2,0)}(t, l)/\Theta^{(0,2)}(t, l) < 0, \tag{58.29}$$

that is, if the second derivatives of $\Theta(t,l)$ have differing signs, then at certain values $T = T_0$ and $L = L_0$ the bias may be equal to zero. In fact, when $S_\Theta(t,l) = 0$, using (58.26) we get

$$\Theta^{(2,0)}(t, l)\,T_0^2 + \Theta^{(0,2)}(t, l)\,L_0^2 = 0,$$

and consequently

$$\frac{T_0}{L_0} = \sqrt{-\frac{\Theta^{(0,2)}(t, l)}{\Theta^{(2,0)}(t, l)}}. \tag{58.30}$$

If the condition (58.28) is satisfied, then the relationship (58.30) is not meaningful. But if the condition (58.29) is valid, then we can determine those values T_0 and L_0 at which $S_\Theta(t,l) = 0$.

We note that the minimum mean-square total error of statistical measurement, $\rho_\Theta(t,l)$, depends on the number of representations N used, in which case the greater this number is, the smaller the value min $\rho_\Theta(t,l)$ can be made. Such a result is natural, since the precision of a statistical measurement is defined in the end by the volume of selective data -- that is, by the number of representations N used.

The processing of just one selective representation, where $N = 1$, is typical of sonar. In this case, as follows from the relationship (58.27) and given the assumptions adopted, min $\rho_\Theta(t,l)$ depends on the characteristics $d_\vartheta(t)$, $d_\vartheta(l)$, τ_ϑ, l_ϑ, $\Theta^{(2,0)}(t,l)$, and $\Theta^{(0,2)}(t,l)$ of the process being investigated. These characteristics can be interpreted as indices of the random process when it is being classified.

§59. Optimization of the Parameters of the Statistical Measuring System

The analysis made above on expression (58.27) demonstrates that $\rho_\Theta(t,l)$, the mean square of the total error, has minimums at certain values of the parameters T and L. This enables us to raise the question of optimizing the statistical measurement with respect to the following criterion:

$$\min_{T,\,L} \rho_\Theta(t,\,l). \tag{59.1}$$

Optimum values, opt T and opt L, correspond to solutions of the following system of equations:

$$\frac{\partial}{\partial T}\rho_\Theta(t,\,l)=0,\quad \frac{\partial}{\partial L}\rho_\Theta(t,\,l)=0. \tag{59.2}$$

Substituting expression (58.27) into (59.2) and differentiating, we get

$$-\frac{4d_\vartheta^{1/2}(t)\,d_\vartheta^{1/2}(l)\,\tau_\vartheta l_\vartheta}{NT^2L}+\left[\frac{\Theta^{(2,0)}(t,\,l)}{24}T^2+\frac{\Theta^{(0,2)}(t,\,l)}{24}L^2\right]\frac{\Theta^{(2,0)}(t,\,l)}{6}T=0,$$

$$(59.3)$$

$$-\frac{4d_\vartheta^{1/2}(t)\,d_\vartheta^{1/2}(l)\,\tau_\vartheta l_\vartheta}{NTL^2}+\left[\frac{\Theta^{(2,0)}(t,\,l)}{24}T^2+\frac{\Theta^{(0,2)}(t,\,l)}{24}L^2\right]\frac{\Theta^{(0,2)}(t,\,l)}{6}L=0.$$

After simple transformations equations (59.3) reduce to the following system:

$$[\Theta^{(2,0)}(t, l)]^2 NT^5L + \Theta^{(2,0)}(t, l)\,\Theta^{(0,2)}(t, l)\,NT^3L^3 = 24^2 d_\vartheta^{1/2}(t)\,d_\vartheta^{1/2}(l)\,\tau_\vartheta l_\vartheta;$$

$$\Theta^{(2,0)}(t, l)\,\Theta^{(0,2)}(t, l)\,NT^3L^3 + [\Theta^{(0,2)}(t, l)]^2 NTL^5 = 24^2 d_\vartheta^{1/2}(t)\,d_\vartheta^{1/2}(l)\,\tau_\vartheta l_\vartheta. \quad (59.4)$$

Subtracting one equation from the other in (59.4), we get

$$[\Theta^{(2,0)}(t, l)]^2 T^4 = [\Theta^{(0,2)}(t, l)]^2 L^4,$$

whence

$$T = L\sqrt{\frac{\Theta^{(0,2)}(t, l)}{\Theta^{(2,0)}(t, l)}}. \quad (59.5)$$

Substituting (59.5) into (59.4) and solving the equations obtained with respect to T and L, we find the following expressions for the optimum parameters, opt T and opt L, of the statistical measuring system:

$$\operatorname{opt} T = \sqrt[6]{\frac{288\, d_\vartheta^{1/2}(t)\, d_\vartheta^{1/2}(l)\,\tau_\vartheta l_\vartheta}{N}\sqrt{\frac{\Theta^{(0,2)}(t, l)}{[\Theta^{(2,0)}(t, l)]^5}}}; \quad (59.6)$$

$$\operatorname{opt} L = \sqrt[6]{\frac{288\, d_\vartheta^{1/2}(t)\, d_\vartheta^{1/2}(l)\,\tau_\vartheta l_\vartheta}{N}\sqrt{\frac{\Theta^{(2,0)}(t, l)}{[\Theta^{(0,2)}(t, l)]^5}}}. \quad (59.7)$$

As can be seen from the relationships (59.6) and (59.7), the optimum smoothing time, opt T, and resolution, opt L, depend on the actual value of the characteristic $\Theta(t, l)$ being measured, as well as on the characteristics $d_\vartheta(t)$, $d_\vartheta(l)$, τ_ϑ, and l_ϑ. Here we encounter the so-called "*a priori* difficulty" in statistical measurement, the essence of which is as follows. If we are to optimize a statistical measurement, we must have *a priori* knowledge of the form of the characteristic being investigated -- that is, its true value. However, the reason the measurements are being made is to obtain an estimate of this characteristic. Thus on the one hand, proper conduct and optimization of a statistical measurement require knowledge of the true value of the probability characteristic being investigated, while, on the other hand, this characteristic cannot be known, in view of the nature of the measuring procedure itself.

Let us discuss some possible ways to overcome this difficulty. The first possibility rests with the fact that to describe the true probability characteristic $\Theta(t,l)$ and the characteristics $d_\vartheta(t)$, $d_\vartheta(l)$, τ_ϑ, and l_ϑ we use their values $\Theta(t,l|m)$, $d_\vartheta(t|m)$, $d_\vartheta(l|m)$, $\tau_\vartheta(m)$, and $l_\vartheta(m)$, which are computed in adopting model m of the process being investigated (this has been treated in Chapter II, where we have discussed the organization and interpretation of experimental research). Naturally, in this case, optimization of the measuring system's parameters is conditional, and in the general case, opt $T(m)$ and opt $L(m)$ do not match the true values, opt T and opt L. Nevertheless, when conducting experiments, our search for basic model m_0 should provide us with values opt $T(m_0)$ and opt $L(m_0)$ which are practically close to the optimum.

Another possibility for overcoming this *a priori* difficulty is the application of a procedure for adaptively optimizing the statistical measurements. This will be discussed in the next paragraph.

Finally, we have yet another possibility for decreasing statistical measurement error, based on the fact that after we make our measurements we introduce *a posteriori* corrections to compensate for estimation bias. Introduction of such corrections requires the solution of corresponding inverse problems, which do not always fit the situation (see Chapter IX). Moreover, such a method does not always produce a reduction in the fluctuation error of the statistical measurement.

§60. Adaptive Optimization*

The principal idea behind adaptive optimization of a statistical measuring system's parameters is as follows: The measuring procedure is broken up into many stages, in which case current information obtained at previous stages is used to optimize the parameters at each successive stage of computing the statistical estimate. In all, the measuring procedure is separated into three basic stages:

---

*Of relevance here is the approach used in [61]. (D.M.)

1. selection of the parameters of the statistical measuring system on the basis of the particular model of the random process adopted;

2. acquisition of initial statistical estimates of the probability characteristics;

3. adaptive optimization of the parameters of the statistical measuring system.

In the first stage (prior to making measurements) the experimenter, who possesses information about possible models of the process, selects the most plausible model. Some aspects of this task are examined in [3], where ways are studied for using a probability model of a nonstationary random process in adaptively optimizing statistical measurement procedures. Next, the investigator determines values for parameters T_1 and L_1 that are optimum for the adopted model m -- that is,

$$T_1 = \text{opt}\ T(m), \quad L_1 = \text{opt}\ L(m). \tag{60.1}$$

Such selection of values for T_1 and L_1, which is considerably subjective, is justified in the first stage, since the experimenter has practically no other possibilities for more justified selection of these parameters.

Of course, a most unfavorable situation is possible in which no model of the process being studied exists at all. In this case, the approach should be as follows: The total time the execution t exists and the range of possible values of parameter l is divided into a certain number of equal intervals ΔT and ΔL. These parameters are adopted as the initial T_1 and L_1. In fact, however, *a priori* information and a model describing the process under investigation (such a model can be unformalized or partially formalized) are always available in experimental sonar research. Therefore, it is always possible to select the initial values T_1 and L_1.

Estimates

$$\hat{\Theta}(t, l),\ \hat{\Theta}^{(2,0)}(t, l),\ \hat{\Theta}^{(0,2)}(t, l),\ \hat{d}_\vartheta(t),\ \hat{d}_\vartheta(l),\ \hat{\tau}_\vartheta,\ \hat{l}_\vartheta \tag{60.2}$$

are determined with the measuring system in the second stage. It is obvious that we cannot expect maximum accuracy in the measurement of these estimates since they were obtained for values of the parameters T_1 and L_1 of the measuring system selected in the first stage, and these values are not optimum with respect to the criterion (59.1).

The third stage in the statistical measurement procedure consists of adaptive optimization. Replacement of the true characteristics

$$\Theta(t, l),\ \Theta^{(2,0)}(t, l),\ \Theta^{(0,2)}(t, l),\ d_\vartheta(t),\ d_\vartheta(l),\ \tau_\vartheta,\ l_\vartheta \tag{60.3}$$

by their statistical estimates (60.2) turns out to be an important step which affects the overall nature of the optimization. This is precisely what enables us to use generated current information about the probability properties of the process for optimization purposes.

Thus (we emphasize once more), adaptive optimization of statistical measurement consists of a multistaged treatment of the process under investigation, using all available information about the characteristics of the random process under investigation at each of these stages. In accordance with the above, at each $(k + 1)$th step in the optimization the values of the parameters $\hat{T}_{k+1}$ and $\hat{L}_{k+1}$ must be found by relationships (59.6) and (59.7), using the values

$$\hat{\Theta}_k^{(2,0)}(t, l),\ \hat{\Theta}^{(0,2)}(t, l),\ \hat{d}_{\vartheta k}(t),\ \hat{d}_{\vartheta k}(l),\ \hat{\tau}_{\vartheta k},\ \hat{l}_{\vartheta k},$$

that is, using the estimates (60.2) obtained at the kth step of statistical measurement:

$$\hat{T}_{k+1} = \sqrt[6]{\frac{288\hat{d}_{\vartheta k}^{1/2}(t)\hat{d}_{\vartheta k}^{1/2}(l)\,\hat{\tau}_{\vartheta k}\hat{l}_{\vartheta k}}{N}\sqrt{\frac{\hat{\Theta}_k^{(0,2)}(t, l)}{\left[\hat{\Theta}_k^{(2,0)}(t, l)\right]^5}}}\,; \tag{60.4}$$

$$\hat{L}_{k+1} = \sqrt[6]{\frac{288\hat{d}_{\vartheta k}^{1/2}(t)\hat{d}_{\vartheta k}^{1/2}(l)\,\hat{\tau}_{\vartheta k}\hat{l}_{\vartheta k}}{N}\sqrt{\frac{\hat{\Theta}_k^{(2,0)}(t, l)}{\left[\hat{\Theta}_k^{(0,2)}(t, l)\right]^5}}}\,. \tag{60.5}$$

The statistical measurement procedure concludes when the values of the parameters $\hat{T}_m$ and $\hat{L}_m$ begin to fluctuate about some fixed values and, as we would expect, when they are close to the optima, opt T and opt L. As an example, satisfaction of the inequalities

$$\frac{1}{k}\sum_{m=M}^{M+k}|\hat{T}_{m+1}-\hat{T}_m|<\varepsilon_T$$

and

$$\frac{1}{k}\sum_{m=M}^{M+k}|\hat{L}_{m+1}-\hat{L}_m|<\varepsilon_L$$

can serve as a criterion for concluding the adaptive optimization procedure: in this case, the values of M, k, ε_T, and ε_L are selected on the basis of the expected potential accuracy of the statistical measurement.

Adaptive optimization involves a greater amount of treatment of the process under investigation. This is fully natural, since increasing the accuracy of a statistical measurement must involve additional costs.

In conclusion, we note that the convergence of the adaptive optimization of measurements is an important factor affecting the effectiveness of this procedure. Some aspects of this problem have been examined in [3], but general conditions for convergence have not been found, unfortunately. Apparently such conditions could be obtained for specific cases of statistical measurement in which not only the class of the random process under investigation, but also the form of characteristic $\Theta(t, l)$ and the model of the process, are defined.

CHAPTER IX

INVERSE PROBABILITY PROBLEMS IN SONAR

§61. Direct and Inverse Problems

In §2, where we examined the sonar model, the echo signal $S(t)$ and the reverberation $F(t)$ at the point of observation were represented as corresponding transformations of the emitted signal $C(t)$:

$$S(t) = M_2 T M_1 A C(t), \qquad (61.1)$$

$$F(t) = m_2 s m_1 A C(t), \qquad (61.2)$$

where A is the operator for the emitting antenna, M_1, M_2, m_1, and m_2 are operators characterizing the effect of the conditions in which a signal propagates in the water medium, and T and s are operators representing the object of detection and the scatterers, respectively.

Expressions (61.1) and (61.2) permit us to define the five following varieties of objects of sonar research:

(a) the water medium (operators M_1, M_2, m_1, and m_2);

(b) the object of detection (operator T);

(c) scatterers (operator s);

(d) the echo signal (function $S(t)$);

(e) the reverberation signal (function $F(t)$).

We introduce the probability characteristics (i.e., statistics) $\Theta(l)$, by which we define the properties of the object of research. Here we designate the probability characteristics of the parameters of the water medium by $\Theta_M(l)$ and $\Theta_m(l)$, those of the objects of detection by $\Theta_T(l)$, the characteristics of echo signals by $\Theta_S(l)$, the characteristics of the scatterers by $\Theta_s(l)$, and the characteristics of

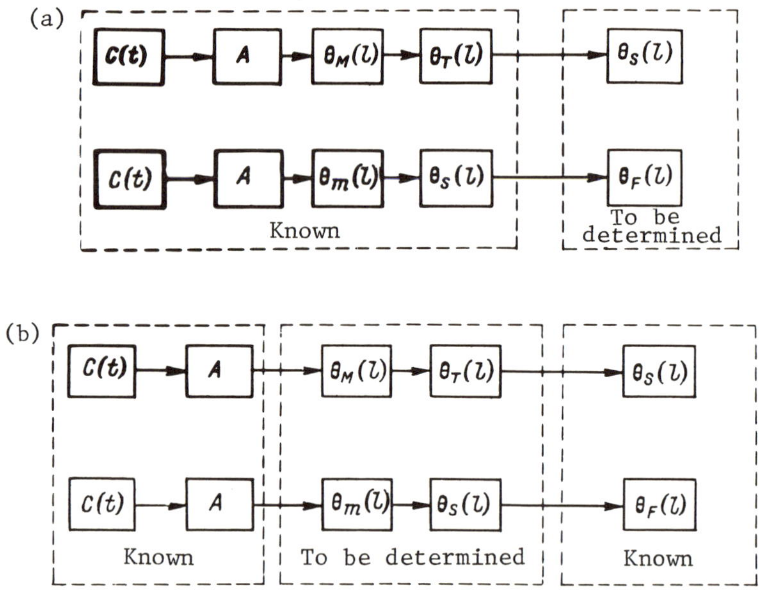

Figure 16. Schema of direct (a) and inverse (b) probability problems in sonar

the reverberation signals by $\Theta_F(l)$. Figure 16 illustrates the schema of direct and inverse probability problems in sonar.

Direct problems are defined as those of determining the probability characteristics of echo signals, $\Theta_S(l)$, and reverberation signals, $\Theta_F(l)$, on the basis of given characterisitcs of the water medium, $\Theta_M(l)$ and $\Theta_m(l)$, the objects of detection $\Theta_T(l)$, and scatterers $\Theta_s(l)$.

The probability models of sonar signals examined in Chapters III, V, and VI permit us to set up the following equations on adequate grounds:

(a) for echo signal characteristics:

$$\Theta_S(l) = M_S[\Theta_M(l),\ \Theta_T(l)]; \tag{61.3}$$

(b) for reverberation signal characteristics:

$$\Theta_F(l) = M_F[\Theta_m(l), \Theta_s(l)]. \qquad (61.4)$$

In this case, operators M_S and M_F are as a matter of course the operators for solving direct probability problems in sonar.

We define inverse problems as those involving the determination of the probability characteristics of the water medium, $\Theta_M(l)$ and $\Theta_m(l)$, or of objects of detection $\Theta_T(l)$, or of scatterers $\Theta_s(l)$, on the basis of given probability characteristics of the echo signals $\Theta_S(l)$, or the reverberation $\Theta_F(l)$. The main feature in solving inverse problems is obtaining inverse equations of the following types from the direct equations (61.3) and (61.4):

(a) for the water medium and objects of detection:

$$\Theta_M(l) = M_{SM}^{-1}[\Theta_S(l), \Theta_T(l)]; \qquad (61.5)$$

$$\Theta_T(l) = M_{ST}^{-1}[\Theta_M(l), \Theta_S(l)]; \qquad (61.6)$$

(b) for the water medium and scatterers:

$$\Theta_m(l) = M_{Fm}^{-1}[\Theta_F(l), \Theta_s(l)]; \qquad (61.7)$$

$$\Theta_s(l) = M_{Fs}^{-1}[\Theta_m(l), \Theta_F(l)]. \qquad (61.8)$$

In this case the operators M_{SM}^{-1}, M_{ST}^{-1}, M_{Fm}^{-1}, and M_{Fs}^{-1} are in some sense inverse to the operators M_S and M_F in the initial (direct) equations (61.3) and (61.4).

Thus, inverse sonar problems are set up to determine various characteristics of the objects of research -- the water medium, objects of detection, and scatterers -- on the basis of known characteristics of echo signals and reverberation signals.

In studying the properties of the water medium, its irregularities, and objects of detection by sonar methods, in most cases the work reduces to solving the examined inverse probability problems [15,17,19,21,30,35], since in this way we gain the possibility of studying the properties of objects of research on the basis of the characteristics of sonar information. Naturally, we can define the characteristics of sonar information (echo signals and reverberation) as actual initial data only after conducting experiments. This means that to obtain initial data we must employ statistical measurements and, consequently, instead of probability characteristics $\Theta_S(l)$ and $\Theta_F(l)$ we have their statistical estimates $\hat{\Theta}_S(l)$ and $\hat{\Theta}_F(l)$. Below, we will discuss the so-called "correctness conditions," which are important in solving inverse sonar problems. In particular, these conditions take account of inaccuracies in the initial data.

§62. Correctness Conditions

An important stage in setting up and solving inverse probability problems in sonar is formulating the conditions for the correctness of these problems. We formulate *correctness conditions* (as defined by Tikhonov) for the following objects when solving inverse problems of mathematical physics: (a) *the class of the initial data* $\mathscr{L}$, and (b) *the correctness set* $\mathscr{F}$.

Determinate problems are usually considered in papers dealing with inverse problems (see, for example, [10]). However, many formulations in the methods which have been developed relate to inverse probability problems as well, in which not the fields or processes themselves, but rather their probability characteristics, are examined. We will discuss below correctness conditions for solving inverse problems applicable to research on the probability characteristics of sonar signals.

The class of initial data $\mathscr{L}$ is normally defined as a class of functions which are given for the problem to be solved. For problems examined here, such functions include $\Theta_S(l)$ and $\Theta_T(l)$ in equation (61.5), $\Theta_M(l)$ and $\Theta_S(l)$, in equation (61.6), $\Theta_F(l)$ and $\Theta_s(l)$ in equation (61.7), and $\Theta_m(l)$ and $\Theta_F(l)$ in equation (61.8). The enumerated functions can be obtained either by direct statistical measurement or as initial data through indirect measurement; (in the latter case, the properties of the data class $\mathscr{L}$ are defined with consideration for recomputation algorithms in the indirect measurement).

The correctness set $\mathscr{F}$ is defined as the set of functions found by solving an inverse problem. In accordance with equations (61.5)-(61.8), such functions are $\Theta_M(l)$, $\Theta_T(l)$, $\Theta_m(l)$, or $\Theta_s(l)$.

An inverse problem is thought to be posed correctly if the following three conditions are satisfied:

(1) *a solution exists;* (in accordance with this condition we consider it given *a priori* that a solution to the problem exists for a certain class of initial data $\mathscr{L}$ and belongs to the solution set $\mathscr{F}$);

(2) *the solution is unique;* (this means that there is only one solution in set $\mathscr{F}$ for a class of initial data $\mathscr{L}$);

(3) *the solution is stable;* (here we have in mind that small variations in the solution correspond to small variations in the initial data $\mathscr{L}$, which do not take the solution beyond the bounds of set $\mathscr{F}$).

Let us examine the conditions formulated above in application to the inverse probability problems of sonar.

The first condition -- existence of a solution -- always holds. In fact when a problem is set up directly, equations (61.3) and (61.4) are derived on the assumption that the physically achievable functions $\Theta_M(l)$, $\Theta_T(l)$, and $\Theta_m(l)$, $\Theta_s(l)$, which are elements of set $\mathscr{F}$, correspond to functions $\Theta_S(l)$ and $\Theta_F(l)$, which are elements of a class of initial data $\mathscr{L}$. The probability characteristics of echo signals, or reverberation, are computed on the basis of physically justified characteristics of the water medium, objects of detection, and scatterers. As a matter of course, this is the essence of developing probability models of sonar information; (such models have been studied in Chapters III-VI).

The second condition -- uniqueness of the solution -- must be proven for each specific case, with a consideration for the form of the operators M_S and M_F as well as for the needed level of probability description of the objects of research. The following concepts, related to the uniqueness of a solution, are presented as part of our general discussion of the correctness of inverse problems.

Selection of the degree to which an echo signal $\Theta_S(l)$, or reverberation signal $\Theta_F(l)$, is described in terms of probability is an important factor governing uniqueness. We know that an n-dimensional probability density $W_n(x, l)$, where

$n \to \infty$, provides a complete probability description of the random process $x(t)$. However, as a rule, it is extremely difficult to obtain an n-dimensional probability density (when $n > 1$) for sonar signals and to conduct experiments. For this reason, signals are described in practice by moments or semi-invariants of different orders.

At least four levels of a probability description can be noted for sonar information:

(1) the energy level: (the mean and variance of the process are determined);

(2) the level of moments or semi-invariants (up through the fourth order) of one-dimensional distributions;

(3) the level of one-dimensional distribution laws or characteristic functions;

(4) the level of correlation functions or energy spectra.

The levels enumerated above provide different degrees of the completeness of the probability description of sonar information, and in the end they determine whether or not the solution of a particular inverse problem is unique.

Obviously, if the sonar signals have a Gaussian probability distribution, then correlation functions provide a complete probability description for them and, consequently, the level of correlation moments (or energy spectra) is adequate to define the uniqueness of solutions to inverse problems. But if the probability distribution of the signals is not Gaussian, then the condition for solution uniqueness may require the use of moments of orders higher than correlation functions to describe the signals.* The order must be selected with a consideration for the form of the operators M_S and M_F.

Another factor affecting the uniqueness of the solution is correspondence between the adopted probability model

---

*In fact, the whole first-order probability density is usually necessary here [52,70]. (D.M.)

and the true object of research. Such correspondence is defined by the operators M_S and M_F in the initial (direct) equations (61.3) and (61.4). Naturally, if the difference between the adopted model and the object of research exceeds a particular value, then the inverse problem would not necessarily have just one solution. Moreover, a solution to a problem which corresponds to the true characteristics of the object of research may, in the case examined here, exist outside of the correctness set $\mathcal{F}$.

The third condition -- stability of the solution -- characterizes the degree to which variations in the initial data affect variations in the solution. The practical significance of this condition is as follows: The initial data -- probability characteristics $\Theta_S(l)$ or $\Theta_F(l)$ -- can be obtained as statistical estimates, $\hat{\Theta}_S(l)$ or $\hat{\Theta}_F(l)$, through experimental research. For a number of reasons these estimates will differ from the probability characteristics noted above. Hence, it is clear that if small differences between characteristics $\Theta_S(l)$ and its estimate $\hat{\Theta}_S(l)$, or between characteristic $\Theta_F(l)$ and its estimate $\hat{\Theta}_F(l)$, cause significant changes in the solution such as to take it out of the bounds of the correctness set $\mathcal{F}$, then we cannot consider the solution to the inverse problem to be stable. The reasons for variations in the initial data can include statistical measurement errors connected with errors in representing the signals, with an inadequacy of the direct model with respect to the object of research, or inaccuracies in the execution of the algorithms for computing statistical estimates, and equipment errors. We have examined these problems in Chapter VIII in our discussion of statistical measurement errors. From this it is clear that achieving stability of the solution reduces as a matter of course to the proper selection of tolerances in the statistical measurement error (of course, in the sense of selecting a certain criterion for the differences between probability characteristics and their statistical estimates).

We note, in conclusion, that we can demonstrate the conditions of the solution's uniqueness and stability only after having defined the specific forms of the operators M_S and M_F in the direct equations (61.3) and (61.4).

In subsequent paragraphs of this chapter we will examine a number of specific inverse probability problems of sonar, some of which have been discussed in [2,14-17,19,21].

§63. Determination of the Probability Density for the Rates of Movement of Scatterers*

Let us examine the following model for the scattering of acoustic waves from discrete irregularities in the water medium. Let the scatterers move with random relative velocities v, which are constant within the limits of the effective duration of the emitted signals. Then, each elementary scattered signal is shifted in frequency by the amount†

$$\Delta\omega = \frac{2\omega_0}{c} v, \tag{63.1}$$

where ω_0 is the central frequency in the emitted signal spectrum and c is the speed of propagation of the acoustic wavefront.

Thus, we reduce the problem to determining the characteristics of marine reverberation in a case where elementary scattered signals depend on the random parameter $\Delta\omega$. If we assume that the probability distribution for the instantaneous values of reverberation is Gaussian (this hypothesis must be tested, of course), then the correlation moment is an adequate level for describing reverberation in terms of probability. For this reason we will use the results of §43 and obtain the necessary initial equation to solve this inverse problem.

The relationship (43.1) permits us to define the autocorrelation function of marine reverberation subject to the fact that the elementary scattered signals creating reverberation depend on random parameters $\boldsymbol{\varepsilon}$. In this case $\boldsymbol{\varepsilon} = \Delta\omega$ -- that is, $\boldsymbol{\varepsilon}$ has one component, and we can write

---

*Here, and in §64-68, recall the basic assumptions of omnidirectionality of the transmitting and receiving apertures (arrays), cf. (*d*), §36, as well as the narrowband nature of the emitted and received signals (*vis-à-vis* the carrier frequency f_o. See, also, [57,58,62] for similar applications. (D.M.)

†Clearly, a monostatic configuration of source and medium is assumed. (D.M.)

$$Z_C(t, \Delta\omega) = Z_C(t) \exp(-j\,\Delta\omega t) \tag{63.2}$$

for the complex envelope of the signal $Z_C(t,\varepsilon) = Z_C(t,\Delta\omega)$. Then the relationship (43.1) for the current-t autocorrelation function of reverberation, $B_F(t,\tau)$, assumes the following form, when (63.2) is taken into account:

$$B_F(t, \tau) = \langle\alpha^2\rangle\, n(t)\, \varphi^2(t) \int_{-\infty}^{\infty}\int_{-\infty}^{\infty} Z_C(t - \tau/2) \times$$

$$\times Z_C^*(t + \tau/2) \exp(j\,\Delta\omega\tau)\, W_{\Delta\omega}(\Delta\omega)\, dt\, d\Delta\omega, \tag{63.3}$$

where $W_{\Delta\omega}(\Delta\omega)$ is the probability density of the random frequency shifts $\Delta\omega$ in the elementary scattered signals. We designate by

$$B_{F0}(t, \tau) = \langle\alpha^2\rangle\, n(t)\, \varphi^2(t) \int_{-\infty}^{\infty} Z_C(t - \tau/2)\, Z_C^*(t + \tau/2)\, dt \tag{63.4}$$

the autocorrelation function of reverberation in the case where the scatterers are not moving. In addition, we consider that

$$\Theta_{\Delta\omega}(\tau) = \int_{-\infty}^{\infty} \exp(j\,\Delta\omega\tau)\, W_{\Delta\omega}(\Delta\omega)\, d\,\Delta\omega \tag{63.5}$$

is the characteristic function of the frequency shifts $\Delta\omega$. Then, on the basis of (63.3)-(63.5), we have

$$B(t, \tau) = B_{F0}(t, \tau)\, \Theta_{\Delta\omega}(\tau). \tag{63.6}$$

From this we find the characteristic function

$$\Theta_{\Delta\omega}(\tau) = \frac{B_F(t, \tau)}{B_{F0}(t, \tau)} \tag{63.7}$$

and the probability density

$$W_{\Delta\omega}(\Delta\omega) = \frac{1}{2\pi}\int_{-\infty}^{\infty} \exp(-j\,\Delta\omega\tau)\,\Theta_{\Delta\omega}(\tau)\,d\tau, \qquad (63.8)$$

which is found to be equal to

$$W_{\Delta\omega}(\Delta\omega) = \frac{1}{2\pi}\int_{-\infty}^{\infty} \exp(-j\,\Delta\omega\tau)\,\frac{B_F(t,\tau)}{B_{F0}(t,\tau)}\,d\tau. \qquad (63.9)$$

Obviously, we can convert from the probability density $W_{\Delta\omega}(\Delta\omega)$ to the probability density $W_v(V)$ for the random rates of movement v of the scatterers. Taking account of (63.1) we get

$$W_v(V) = \frac{c}{2\omega_0} W_{\Delta\omega}\left(\frac{2\omega_0}{c}\right) \qquad (63.10)$$

and, consequently, after determining the probability density $W_{\Delta\omega}(\Delta\omega)$ with the assistance of (63.9), we can find $W_v(V)$ using (63.10).

In accordance with (63.9) the autocorrelation function of reverberation, $B_F(t,\tau)$, which is obtained experimentally by statistical measurement, is the initial characteristic we use in solving this inverse problem. The autocorrelation function $B_{F0}(t,\tau)$ is evaluated from the emitted signal's complex envelope $Z_C(t)$, using (63.4).

It is important (from the standpoint of the conditions for the correctness of the solution to this inverse problem) to turn our attention to the two hypotheses assumed above. The first hypothesis involves the constancy of the scatterer's rate of movement over the duration of the emitted signal: The initial relationship (63.2) is written on this basis. The second hypothesis states that the probability distribution of instantaneous values of reverberation is Gaussian. Both hypotheses must be checked out and proven. If they are not entirely true, the uniqueness and stability of the solution may be affected. We note that in this case we are discussing the correspondence between the adopted model and the object of research. As had been indicated in Chapter II, establishing this correspondence is one of the tasks of experimental research and statistical measurement (these hypotheses are satisfied if we can prove that the model adopted above is the base model).

§64. Determination of the Spectrum of the Fluctuations in the Rate of Movement of Scatterers

In solving the inverse problem examined in §63, where we determined the probability density of the rates of movement of the scatterers, we assumed that the scatterers move uniformly and linearly, with random velocities, during the effective duration of the emitted signal. Such a model for scatterer movement does not always hold, and if we test such a hypothesis we may find that it would have to be rejected. Therefore, the next step in making the model of scatterer movement more sophisticated is to reject the hypothesis that the scatterers move uniformly and linearly.

We will assume that the scatterers move along random trajectories, and that the random Doppler frequency shifts $\Delta\omega(t)$ in the elementary scattered signals depend on time* and can be represented as:

$$\Delta\omega(t) = \frac{2\omega_0}{c} v(t), \tag{64.1}$$

where ω_0 is the central frequency of the emitted signal's spectrum, c is the rate of propagation of the acoustic wave, and $v(t)$ is a random radial component of scatterer velocity.

Let us return to the description of marine reverberation at the correlation level. We assume, as previously, that the instantaneous values of reverberation have a Gaussian probability distribution. Consequently, we must use the results in §43, where we obtained an expression for the current-t correlation function for the case in which ele-

---

*Cf. comments in the Introduction, and the procedures, generally of [65]; here the Doppler effect depends on both space and time [65]. (D.M.)

mentary scattered signals have random form. We use expression (43.1) for this purpose, rewriting it as

$$B_F(t, \tau) = \langle \alpha^2 \rangle n(t) \varphi^2(t) \times$$
$$\times \int_{-\infty}^{\infty} \langle Z_C [t - \tau/2, \varepsilon(t - \tau/2)] Z_C^* [t + \tau/2, \varepsilon(t + \tau/2)] \rangle dt, \quad (64.2)$$

where statistical averaging of the integrand is performed with respect to the following function $\varepsilon(t)$, interpreted as the instantaneous phase of the elementary scattered signals:

$$\varepsilon(t) = \int_0^t \Delta\omega(t')\, dt' = \frac{2\omega_0}{c} \int_0^{t'} v(t')\, dt. \quad (64.3)$$

We note that in this case $\varepsilon(t)$ is a random process, whereas in the discussion above the form of the elementary scattered signals depend on the random parameter ε. Such a conversion to a random process follows naturally when the results presented in §43 are generalized.

In a way similar to that employed for (63.2), we present the complex envelope $Z_C[t, \varepsilon(t)]$ of the signal in the form

$$Z_C [t, \varepsilon(t)] = Z_C(t) \exp[-j\varepsilon(t)]$$

or, taking account of (64.3), as

$$Z_C[t, \varepsilon(t)] = Z_C(t) \exp\left[-j \frac{2\omega_0}{c} \int_0^t v(t')\, dt'\right]. \quad (64.4)$$

Now let us examine the average value of the product

$$I(t, \tau) = \langle Z_C [t - \tau/2, \varepsilon(t - \tau/2)] Z_C^* [t + \tau/2, \varepsilon(t + \tau/2)] \rangle \quad (64.5)$$

in the integrand of (64.2). Substituting (64.4) into (64.5), we get

$$I(t,\ \tau) = Z_C(t-\tau/2)\, Z_C^*(t+\tau/2) \times$$
$$\times \left\langle \exp\left[-j\frac{2\omega_0}{c}\int_0^{t-\tau/2} v(t')\,dt' + j\frac{2\omega_0}{c}\int_0^{t+\tau/2} v(t')\,dt'\right]\right\rangle. \qquad (64.6)$$

We designate by

$$\Psi(t) = \frac{2\omega_0}{c}\int_0^t v(t')\,dt' \qquad (64.7)$$

a function characterizing the instantaneous phase of the elementary scattered signals. Then, considering that

$$\Theta_\Psi(\eta_1,\ \eta_2;\ t_1,\ t_2) = \langle \exp\,[j\eta_1\Psi(t_1) + j\eta_2\Psi(t_2)]\rangle \qquad (64.8)$$

is a two-dimensional characteristic function of the random process $\Psi(t)$, we have

$$\Theta_\Psi(-1,\ 1;\ t-\tau/2,\ t+\tau/2) = \langle \exp\,[-j\Psi(t-\tau/2) + j\Psi(t+\tau/2)]\rangle. \qquad (64.9)$$

Now the relationships (64.6), (64.7), and (64.9) permit us to write $I(t,\tau)$ as follows:

$$I(t,\ \tau) = Z_C(t-\tau/2)\, Z_C^*(t+\tau/2)\, \Theta_\Psi(-1,\ 1;\ t-\tau/2,\ t+\tau/2). \qquad (64.10)$$

We postulate yet another limitation imposed on the model of scatterer movement: We assume that the rate of movement, $v(t)$, is a Gaussian stationary random process. In this case it is known (see, for example, [12]), that

$$\Theta_\Psi(-1,\ 1;\ t_1,\ t_2) = \exp\left[-\frac{d_\Psi(t_1)\,\eta_1^2 + d_\Psi(t_2)\,\eta_2^2 + 2B_\Psi(t_1,\ t_2)\,\eta_1\eta_2}{2}\right], \qquad (64.11)$$

where $d_\Psi(t)$ is the variance of the process $\Psi(t)$ and $B_\Psi(t_1,t_2)$ is its correlation function; cf. (33.11) and (33.12).

We note that the instantanteous phase (64.7) is a nonstationary random process, since it is associated with a stationary process $v(t)$ by a definite integral with a variable upper limit. In view of this, the variance $d_\Psi(t)$ depends on current time (t_1 or t_2) while the autocorrelation function $B_\Psi(t_1,t_2)$ depends on two reading times.

Obviously,

$$\Theta_\Psi(-1,\ 1;\ t-\tau/2,\ t+\tau/2)=$$

$$=\exp\left[-\frac{d_\Psi(t-\tau/2)+d_\Psi(t+\tau/2)-2B_\Psi(t-\tau/2,\ t+\tau/2)}{2}\right] \qquad (64.12)$$

follows from (64.11). The relationship (64.7) permits us to express the statistics in (64.12) by the statistic of the initial random process $v(t)$. Let us consider its variance, d_v, and autocorrelation function, $B_v(\tau)$:

$$d_v = \langle v^2(t)\rangle; \qquad (64.13)$$

$$B_v(\tau) = \langle v(t)\, v(t+\tau)\rangle. \qquad (64.14)$$

From (64.7) and (64.14) we get

$$d_\Psi(t)=\left(\frac{2\omega_0}{c}\right)^2\int_0^t\int_0^t B_v(t'-t'')\,dt'\,dt''; \qquad (64.15)$$

$$B_F(t_1,\ t_2)=\left(\frac{2\omega_0}{c}\right)^2\int_0^{t_1}\int_0^{t_2} B_v(t'-t'')\,dt'\,dt''. \qquad (64.16)$$

Now, we consider the general relationship

$$\int_0^{t_1}\int_0^{t_2} B_v(t'-t'')\,dt'\,dt'' = t_1\int_0^{t_1} B_v(t')\left(1-\frac{t'}{t_1}\right)dt' +$$

$$+t_2\int_0^{t_2} B_v(t')\left(1-\frac{t'}{t_2}\right)dt' - |t_2-t_1|\int_0^{|t_1-t_2|} B_v(t')\left(1-\frac{t'}{|t_2-t_1|}\right)dt', \tag{64.17}$$

which is valid for the case in which $B_v(\tau)$ is even. This condition is satisfied, of course, since $B_v(\tau)$ is the autocorrelation function of a real stationary random process $v(t)$.

Next, making use of (64.15)-(64.17), we get

$$d_\Psi(t-\tau/2) = 2\left(\frac{2\omega_0}{c}\right)^2(t-\tau/2)\int_0^{t+\tau/2} B_v(t')\left(1-\frac{t'}{t-\tau/2}\right)dt'; \tag{64.18}$$

$$d_\Psi(t+\tau/2) = 2\left(\frac{2\omega_0}{c}\right)^2(t+\tau/2)\int_0^{t-\tau/2} B_v(t')\left(1-\frac{t'}{t+\tau/2}\right)dt'; \tag{64.19}$$

$$B_\Psi(t-\tau/2,\ t+\tau/2) = \left(\frac{2\omega_0}{c}\right)^2(t-\tau/2)\int_0^{t-\tau/2} B_v(t')\left(1-\frac{t'}{t-\tau/2}\right)dt' +$$

$$+\left(\frac{2\omega_0}{c}\right)^2(t+\tau/2)\int_0^{t+\tau/2} B_v(t')\left(1-\frac{t'}{t+\tau/2}\right)dt' -$$

$$-2|\tau|\left(\frac{2\omega_0}{c}\right)^2\int_0^{|\tau|} B_v(t')\left(1-\frac{t'}{|\tau|}\right)dt'. \tag{64.20}$$

Now, we substitute the functions $d_\Psi(t+\tau/2)$, $d_\Psi(t-\tau/2)$, and $B_\Psi(t-\tau/2,\ t+\tau/2)$, defined by the relationships (64.18)-(64.20), into (64.12). Then, after simple transformations, we find that

$$\Theta_\Psi(-1,\ 1;\ t-\tau/2,\ t+\tau/2) = \exp\left[-2|\tau|\left(\frac{2\omega_0}{c}\right)^2\int_0^{|\tau|} B_v(t')\left(1-\frac{t'}{|\tau|}\right)dt'\right]. \tag{64.21}$$

The expressions (64.2), (64.5), (64.10), and (64.21) enable us to specify the autocorrelation function of reverberation as

$$B_F(t,\ \tau) = B_{F0}(t,\ \tau)\exp\left[-2|\tau|\left(\frac{2\omega_0}{c}\right)^2 \int\limits_0^{|\tau|} B_v(t')\left(1-\frac{t'}{|\tau|}\right)dt'\right], \tag{64.22}$$

where

$$B_{F0}(t,\ \tau) = \langle\alpha^2\rangle\, n(t)\, \varphi^2(t) \int\limits_{-\infty}^{\infty} Z_C(t-\tau/2)\, Z_C^*(t+\tau/2)\, dt \tag{64.23}$$

is the autocorrelation function of reverberation derived without consideration for the effects of scatterer movement. It follows from (64.22) that

$$2|\tau|\left(\frac{2\omega_0}{c}\right)^2 \int\limits_0^{|\tau|} B_v(t')\left(1-\frac{t'}{|\tau|}\right)dt' = \ln\frac{B_{F0}(t,\ \tau)}{B_F(t,\ \tau)}. \tag{64.24}$$

Relationship (64.24) defines an integral equation in which the unknown is the autocorrelation function $B_v(\tau)$ for fluctuations in scatterer velocities $v(t)$. We add to our examination the energy spectrum, $G_v(\omega)$, of fluctuations in velocities, which is associated with $B_v(\tau)$ by a Fourier transform.*

$$B_v(\tau) = \frac{1}{2\pi}\int\limits_0^{\infty} G_v(\omega)\cos\omega\tau\, d\omega. \tag{64.25}$$

Strictly speaking, in solving this inverse problem it makes no difference as to which characteristic we must define -- the correlation function $B_v(\tau)$ or its spectrum $G_v(\omega)$. Selection of the form of the unknown characteristic is only a matter of convenience in this case. We will subsequently seek the energy spectrum $G_v(\omega)$.

---

*A special case of the familiar Wiener-Khintchine theorem ([12], Sec. 3.2, for example). (D.M.)

By substituting (64.25) into (64.24) we get

$$\frac{|\tau|}{\pi}\left(\frac{2\omega_0}{c}\right)^2 \int_0^{|\tau|}\int_0^{\infty} G_v(\omega)\left(1-\frac{t'}{|\tau|}\right)\cos\omega t'\,d\omega\,dt' = \ln\frac{B_{F0}(t,\tau)}{B_F(t,\tau)}. \quad (64.26)$$

Next, considering that

$$\int_0^{|\tau|}\left(1-\frac{t'}{|\tau|}\right)\cos\omega t'\,dt' = \frac{2\sin^2\frac{\omega|\tau|}{2}}{\omega^2|\tau|}, \quad (64.27)$$

we get

$$\frac{\tau^2}{2\pi}\left(\frac{2\omega_0}{c}\right)^2 \int_0^{\infty} G_v(\omega)\left(\frac{\sin\frac{\omega\tau}{2}}{\omega\tau/2}\right)^2 d\omega = \ln\frac{B_{F0}(t,\tau)}{B_F(t,\tau)} \quad (64.28)$$

from (64.26). As had been indicated above, the energy spectrum $G_v(\omega)$ of the fluctuations in velocities given in (64.28) is the function being sought in our solution of the inverse problem under consideration here. Designating by $K(\omega\tau)$ the kernel of the integral equation (64.28), e.g.,

$$K(\omega\tau) = \left(\frac{\sin\omega\tau/2}{\omega\tau/2}\right)^2, \quad (64.29)$$

and by $E(\tau)$ the initial data of the problem

$$E(\tau) = \frac{2\pi}{\tau^2}\left(\frac{c}{2\omega_0}\right)^2 \ln\frac{B_{F0}(t,\tau)}{B_F(t,\tau)}, \quad (64.30)$$

we have

$$E(\tau) = \int_0^{\infty} G_v(\omega)\,K(\omega,\tau)\,d\omega. \quad (64.31)$$

It is interesting to note that this function $E(\tau)$, defined as in (64.30) does not depend on time t, though it is expressed by the current-t correlation functions of reverberation. The reason for this is the nature of the non-stationarity of reverberation, which in this case is assumed to be reducible to a stationary homogeneous random process [14,17]. It follows from this fact, in particular, that for the model of scatterer movement adopted here, the way the correlation functions $B_F0(t,\tau)$ and $B_F(t,\tau)$ depend on time t is the same.

Here is an important feature of the integral equation (64.31): The kernel of this equation, $K(\omega,\tau)$, is a function of the product of the variables, ω,τ. This means that we can use the Mellin inversion formula to solve equation (64.31) -- that is, to find the spectrum $G_v(\omega)$ [15].

The Mellin inversion formula $\mathscr{A}(s)$ for a certain function $A(x)$ is defined by an integral of the form

$$\mathscr{A}(s) \equiv \int_0^\infty A(x)\, x^{s-1}\, dx. \tag{64.32}$$

A transform inverse to (64.32) has the following form:

$$A(x) = \frac{1}{2\pi j} \int_{c-j\infty}^{c+j\infty} \mathscr{A}(s)\, x^{-s}\, ds. \tag{64.33}$$

To solve equation (64.31) we convert from the function $E(\tau)$ to its Mellin transform

$$\mathscr{E}(s) = \int_0^\infty E(\tau)\, \tau^{s-1}\, d\tau, \tag{64.34}$$

such that

$$\mathscr{E}(s) = \int_0^\infty \int_0^\infty G_v(\omega)\, K(\omega\tau)\, \tau^{s-1}\, d\tau\, d\omega. \tag{64.35}$$

We note that

$$\int_0^\infty K(\omega\tau)\,\tau^{s-1}\,d\tau = \omega^{-s}\,\mathcal{K}(s),$$

$$\mathcal{K}(s) = \int_0^\infty K(x)\,x^{s-1}\,dx \tag{64.36}$$

is a Mellin transform of the integral's kernel, given the substitution $x = \omega\tau$. Then, from (64.35) we convert to the relationship

$$\mathcal{E}(s) = \mathcal{K}(s)\int_0^\infty G_v(\omega)\,\omega^{-s}\,d\omega. \tag{64.37}$$

It follows from the definition of a Mellin transform that

$$\mathcal{G}_v(1-s) = \int_0^\infty G_v(\omega)\,\omega^{-s}\,d\omega. \tag{64.38}$$

Then, from (64.37) and (64.38) we have

$$\mathcal{G}_v(1-s) = \frac{\mathcal{E}(s)}{\mathcal{K}(s)}, \tag{64.39}$$

from which we get the equality

$$\mathcal{G}_v(s) = \frac{\mathcal{E}(1-s)}{\mathcal{K}(1-s)}. \tag{64.40}$$

Finding from (64.40) inverse Mellin transforms of the (64.33) type, we get the desired solution, $G_v(\omega)$, of the integral equation (64.31) in the following form:

$$G_v(\omega) = \frac{1}{2\pi j}\int_{c-j\infty}^{c+j\infty} \frac{\mathcal{E}(1-s)}{\mathcal{K}(1-s)}\,\omega^{-s}\,ds. \tag{64.41}$$

It follows from the solution (64.41) of the inverse problem under consideration here that to find the energy spectrum $G_v(\omega)$ of the fluctuations in scatterer velocity we must find the Mellin transforms $\mathscr{E}(s)$ and $\mathscr{K}(s)$ of the functions $E(\tau)$ and $K(x)$, defined by relationships (64.30) and (64.29) respectively, then replace their arguments by $1 - s$, and finally use the inversion formula (64.33) in accordance with (64.41).

Now let us find the Mellin transform $\mathscr{E}(s)$ and $\mathscr{K}(s)$.

In accordance with (64.30) and (64.34) we have

$$\mathscr{E}(s) = 2\pi\left(\frac{c}{2\omega_0}\right)^2 \int_0^\infty \ln\frac{B_{F0}(t,\tau)}{B_F(t,\tau)}\,\tau^{s-3}\,d\tau, \qquad (64.42)$$

whence

$$\mathscr{E}(1-s) = 2\pi\left(\frac{c}{2\omega_0}\right)^2 \int_0^\infty \ln\frac{B_{F0}(t,\tau)}{B_F(t,\tau)}\,\tau^{-s-2}\,d\tau. \qquad (64.43)$$

Inasmuch as the functions $B_{F0}(t,\tau)$ and $B_F(t,\tau)$ are the initial data of the problem, the function $\mathscr{E}(1 - s)$ can also be thought of as given. Next, we obtain $\mathscr{K}(s)$. In accordance with (64.29) and (64.36), for $\mathscr{K}(s)$ we have

$$\mathscr{K}(s) = \int_0^\infty \left(\frac{\sin\frac{x}{2}}{x/2}\right)^2 x^{s-1}\,dx. \qquad (64.44)$$

We make use of the following tabulated integral:

$$\int_0^\infty \sin^2(au)\,u^{p-1}\,du = \frac{\Gamma(p)\cos\frac{\pi p}{2}}{2^{p+1}a^p}, \qquad \mathrm{Re}(p) < 0, \qquad (64.45)$$

where $\Gamma(p)$ is a gamma-function. Then, it follows from (64.44) and (64.45) that

$$\mathscr{K}(s) = -2\Gamma(s-2)\cos\frac{\pi(s-2)}{2}, \quad 0 < \mathrm{Re}\,S < 2. \qquad (64.46)$$

Replacing argument s by $1 - s$ in (64.46), and also considering that

$$\Gamma(p)\Gamma(-p) = -\frac{\pi}{p \sin(\pi p)},$$

for $\mathcal{K}(1 - s)$, we get

$$\mathcal{K}(1-s) = \frac{\pi}{(s+1)\Gamma(s+1)\sin\frac{\pi(s+1)}{2}}. \qquad (64.47)$$

Now we can write the solution, (64.41), of the integral equation as

$$G_v(\omega) = \frac{1}{2\pi^2 j} \int_{c-j\infty}^{c+j\infty} \mathcal{E}(1-s)\ (s+1)\Gamma(s+1)\sin\frac{\pi(s+1)}{2}\,\omega^{-s}\,ds. \qquad (64.48)$$

And so, the inverse problem we are considering here, of determining the energy spectrum of the fluctuations in scatterer velocities, is solved in accordance with the following algorithm: We compute the correlation function $B_{F0}(t,\tau)$ and measure the correlation function $B_F(t,\tau)$. Next, we determine the function $\mathcal{E}(1 - s)$ using relationship (64.43), and then we determine the desired spectrum $G_v(\omega)$ using (64.48).

It should be kept in mind that in solving this inverse problem we have assumed that the fluctuations in scatterer velocities follow a Gaussian probability distribution. If the validity of this hypothesis has not been demonstrated, then the solution -- the energy spectrum $G_v(\omega)$ -- may not satisfy the conditions of uniqueness and stability. This circumstance must be taken into account when experiments are set up to obtain initial data for the solution of the inverse problem.

§65. Determination of the Relative Level of Coherent Scattering

The following problem is encountered in research regarding the structure of echo signals and the objects of detection, as well as the properties of marine reverberation:

Using measured characteristics of the received signals, we must determine the value of the parameter Q, which characterizes the ratio between the average energy E_C of the signal's coherent component and the average energy $d_X T_{ef}$ of its noncoherent random component or, to put it another way, the ratio between the average intensities I_C and I_X of these components, since $E_C = I_C T_{ef}$, $d_X = I_X$, viz:

$$Q = \frac{E_C}{d_X T_{ef}} = \frac{I_C}{I_X}. \tag{65.1}$$

The parameter Q is a relative measure of specular (i.e., coherent) reflection and the random (i.e., incoherent) scattering of acoustic waves and, consequently, it defines the energy characteristics of echo signals or reverberation.

The first method for determining Q reduces to studying the mathematical expectation of the normalized characteristic of similarity for emitted signals and echo signals. Expression (30.10) for the mathematical expectation $|m_\varkappa(\Omega,\tau)|$ for the characteristic of signal similarity was first derived in §30, where we examined echo signals consisting of an additive mixture of a signal of known form and a random component. It follows from (30.10) that

$$|m_\varkappa(\Omega, \tau)|^2 = \frac{Q}{1+Q} |\chi(\Omega, \tau)|^2, \tag{65.2}$$

where $\chi(\Omega,\tau)$ is the ambiguity function of the emitted signal. From the relationship (65.2) we find that

$$Q = \frac{|m_\varkappa(\Omega, \tau)|^2}{|\chi(\Omega, \tau)|^2 - |m_\varkappa(\Omega, \tau)|^2}. \tag{65.3}$$

Considering that $|\chi(0,0)|^2 = 1$, we get the following particular formula for Q at the point $\Omega = 0$, $\tau = 0$ in the frequency-time domain:

$$Q = \frac{|m_\varkappa(0, 0)|^2}{1 - |m_\varkappa(0, 0)|^2}. \tag{65.4}$$

Thus, to determine the value of the parameter Q, in the general case we must compute the ambiguity function of the emitted signal $\chi(\Omega,\tau)$ and measure the mathematical expectation $|m_{\varkappa}(\Omega,\tau)|$ for the normalized characteristic of similarity for emitted signals and echo signals.

The second method for determining the parameter Q reduces to studying the one-dimensional probability density of the received signal (see §46). Examining, in particular, the probability density $W(E)$ for the envelope of the sum of coherent and random Gaussian components, using the symbols adopted and in accordance with (46.35), we have

$$W(E)=\frac{E}{d_X}\exp\left(-\frac{E^2}{2d_X}-Q\right)I_0\left(\frac{E}{\sqrt{d_X}}\sqrt{Q}\right),\quad E\geqslant 0. \tag{65.5}$$

It is evident from (65.5) that the probability density $W(E)$ contains information about the value of the parameter Q. This information is also contained, naturally, in the moments of the distribution (65.5) and, in particular, in the mathematical expectation

$$m_E=\int_0^\infty EW(E)\,dE. \tag{65.6}$$

Substituting (65.5) into (65.6) produces the relationship

$$m_E=\sqrt{2d_X}\,\Gamma\left(\frac{3}{2}\right){}_1F_1\left(-\frac{1}{2},\ 1,\ -Q\right), \tag{65.7}$$

where $\Gamma(3/2)=\sqrt{\pi}/2$, and ${}_1F_1(-1/2,\ 1,\ -Q)$ is a confluent hypergeometric function.* If we designate by

---

*An equivalent form is ${}_1F_1(-1/2,\ 1,\ -Q) = e^{-Q/2}[(1+Q/2)\times I_0(Q/2)+(Q/2)I_1(Q/2)]$, where I_0, I_1 are modified functions of the first kind (cf. [12] p. 1076, for example). See also [54] pp. 275-282 and [55] for a discussion and tables. (D.M.)

$$ {}_1F_1^{-1}\left(-\frac{1}{2}, 1, \frac{m_E}{\sqrt{2d_X}\,\Gamma\left(\frac{3}{2}\right)}\right) $$

a function inverse to ${}_1F_1(-1/2, 1, -Q)$, then on the basis of (65.7) we get (formally)

$$ Q = -{}_1F_1^{-1}\left(-\frac{1}{2}, 1, \frac{m_E}{\sqrt{2d_X}\,\Gamma\left(\frac{3}{2}\right)}\right). \quad (65.8) $$

In some cases, using the relationship (65.8) may not produce a unique solution to the inverse problem under consideration here, since when the distribution of the random component of the resultant signal is non-Gaussian, formula (65.5) is not valid for the probability density $W(E)$. This fact had been mentioned in §46 and will be discussed in the next paragraph.

It is interesting to note that the first method for determining the parameter Q, described above, does not have this shortcoming, since the mathematical expectation $|m_\varkappa(\Omega,\tau)|$ does not depend on the form of the distribution for the random component of the observed sonar signal, but depends only on its average intensity.

§66. Determination of the Average Number of Scatterers [58,65]

The average number of scattering irregularities per unit domain within a certain volume of the water medium is an important characteristic of such irregularities. Such a "scatterer density" is a characteristic of their physical nature and is viewed as an important index of the scattering irregularities. We can use various methods to determine the average number of scatterers, selecting particular characteristics of the sonar signals received as the initial data. The simplest method for solving such an inverse problem is to determine the average number of elementary scattered signals on the basis of the relative variance of the reverberation. In accordance with (40.9) the variance $d_F(t)$ of the reverberation signal is expressed by

$$d_F(t) = \langle \alpha^2 \rangle\, n(t)\, \varphi^2(t)\, E_C, \tag{66.1}$$

where $\langle \alpha^2 \rangle$ is the second first-order moment of the amplitudes of the elementary scattered signals per unit time or range, $n(t)$ is the unknown function defining the number of elementary signals per unit time or range, $\varphi(t)$ is a function characterizing the drop in signal intensity with respect to time (e.g., range), and E_C is the energy of the emitted signal. At $n(t) = 1$, that is, when a signal from one scatterer arrives at the point of reception, we have

$$d_{F1}(t) = \langle \alpha^2 \rangle\, \varphi^2(t)\, E_C. \tag{66.2}$$

for the variance of the reverberation $d_{F_1}(t)$. It follows from (66.1) and (66.2) that

$$n(t) = \frac{d_F(t)}{d_{F1}(t)}. \tag{66.3}$$

In the relationship (66.3), $d_{F_1}(t)$ must be computed and $d_F(t)$ must be measured. However, this method may be found to be ineffective, because computing $d_{F_1}(t)$ involves adoption of a whole series of hypotheses on the properties of one scatterer. Therefore, we will examine yet another method for determining $n(t)$, based on studying the probability distributions of reverberation signals given a finite number of scatterers.

It had been demonstrated in §46 that the probability distribution of reverberation signals is other than Gaussian when the average number $n(t)$ of elementary scattered signals is small. The coefficient of excess, $\gamma_e(t)$, for the distribution of instantaneous values can serve as a measure of the extent to which the distribution differs from Gaussian. In accordance with (46.11) this coefficient is defined by the relationship

$$\gamma_e(t) = \frac{AC}{n(t)}, \tag{66.4}$$

where

$$A = \frac{\langle \alpha^4 \rangle}{\langle \alpha^2 \rangle^2}, \quad C = \frac{\int\limits_{-\infty}^{\infty} C^4(t)\,dt}{\left[\int\limits_{-\infty}^{\infty} C^2(t)\,dt\right]^2} \qquad (66.5)$$

are parameters whose values are defined by the moments of the distributions of the random amplitudes of the scattered signals and by the form of the emitted signal $C(t)$. It follows from (66.4) that

$$n(t) = \frac{AC}{\gamma_e(t)}. \qquad (66.6)$$

In the relationship (66.6), A and C must be computed and $\gamma_e(t)$ must be measured.

We note that coherent scattering should also affect the value of the coefficinet of excess, if such scattering exists. However, as follows from §65, the relative level Q of the coherent component can be determined by studying the characteristic of similarity for emitted signals and echo signals. The possible effects the average number $n(t)$ of scattered signals may have on the size of $\gamma_e(t)$ and Q have been discussed in §46.

§67. Determination of the Frequency Properties of Scatterers

In our study of wideband signals in §44, we examined the space-time correlation of marine reverberation. In correspondence with the relationship (44.12) we can define the current-t energy spectrum $G_F(t,\omega)$ of reverberation as

$$G_F(t, \omega) = \frac{8\Gamma^2 \langle a^2 \rangle}{\pi c^4} \frac{n(t)}{t^4} |g(\omega)|^2 \exp[-0.2c\beta(\omega)t] \langle |K_p(\omega, \xi)|^2 \rangle, \quad (67.1)$$

where Γ is a parameter taking account of the energy characteristics of the acoustic antennas,* c is the speed of propagation of the acoustic wave, $\langle \alpha^2 \rangle$ is the second initial moment of the amplitude distribution of the elementary scattered signals, $g(\omega)$ is the frequency spectrum of the emitted signal, $\beta(\omega)$ is the frequency characteristic of absorption in the water medium, and $\langle |K_p(\omega,\xi)|^2 \rangle$ is the average energy-frequency characteristic of the scatterers. We designate by

$$A(t) = \frac{8\Gamma^2}{\pi c^4} \frac{\langle a^2 \rangle n(t)}{t^4} \qquad (67.2)$$

a function characterizing the change with respect to time of parameters not depending on frequency. Then, from (67.1) and (67.2) we have

$$G_F(t, \omega) = A(t)\,|g(\omega)|^2 \exp[-0.2c\beta(\omega)t]\,\langle |K_p(\omega, \xi)|^2 \rangle. \qquad (67.3)$$

Equation (67.3) shows that the frequency spectrum of reverberation is defined by three functions: $|g(\omega)|^2$, $\exp[-0.2c\beta(\omega)t]$, and $\langle |K_p(\omega,\xi)|^2 \rangle$, the first of which, (the spectrum of the emitted signal) is known, the second of which (absorption in the water medium) can be evaluated given the physical characteristics of the medium, and the third of which (frequency characteristic of the scatterers) must be found. Thus, if the frequency spectrum of reverberation, $G_F(t,\omega)$, has been measured, then, as follows from (67.3), the unknown frequency characteristic of the scatterers $\langle |K_p(\omega,\xi)|^2 \rangle$ is expressed by known functions:

$$\langle |K_p(\omega, \xi)|^2 \rangle = \frac{G_F(t, \omega)}{A(t)\,|g(\omega)|^2 \exp[-0.2c\beta(\omega)t]}. \qquad (67.4)$$

§68. Determination of the Spatial Distribution of Scatterers

We will examine the following spatial model of scattering irregularities: Let the irregularities be distri-

---

*We recall the assumption of omnidirectionality and frequency insensitivity of these acoustic antennas. See the footnote comment, p. 124. (D.M.)

buted through a layer with thickness h, and let the function $W_\theta(\theta)$ describe the average scatterer density within this layer, where θ is an angle in the vertical plane. The inverse problem in this case involves determining the function $W_\theta(\theta)$ on the basis of the spatial correlation function of marine reverberation.

Such a scattering model can hold when surface reverberation caused by scattering from irregularities in the surface layer (air bubbles) is observed, when reverberation is observed from so-called underwater sound-scattering layers, and in some other cases.

Starting with the general relationship (44.13) for the spatial correlation function $B_F(t,0;\ r_r) = B_F(t,r_r)$, for narrow-band signals we have

$$B_F(t,\ r_r) = P(t) \int\limits_{\theta_1(t)}^{\theta_2(t)} W_\theta(\theta) \cos[\omega_0 \Delta t(r_r,\ \theta)] \cos\theta\, d\theta, \qquad (68.1)$$

where, using the symbols adopted in §44,

$$P(t) = \frac{8\Gamma^2}{\pi c^4} \frac{\langle a^2\rangle n(t)}{t^4} \langle |K_p(\omega_0\xi)|^2 \exp[-0.2c\beta(\omega_0)t]\rangle \int\limits_0^\infty |g(\omega)|^2 d\omega \qquad (68.2)$$

is a function characterizing the dependence of the reverberation intensity on time, $\theta_1(t)$ and $\theta_2(t)$ are angles in the vertical plane tracing the outline of the scattering layer in which the scatterers are distributed, ω_0 is the central frequency of the emitted signal spectrum, and

$$\Delta t(r_r, \theta) = \frac{r_r}{c} \sin\theta. \qquad (68.3)$$

Next, let us examine vertical correlation "along the wavefront" and assume that the angles

$$\theta_1(t) = -\frac{h}{ct}, \quad \theta_2(t) = \frac{h}{ct} \qquad (68.4)$$

satisfy the condition $h/ct \ll 1$, which corresponds to a rather thin layer of scatterers. Then, on the basis of (68.1), (68.3), and (68.4) we have

$$B_F(t, r_r) = P(t) \int_{-\frac{h}{ct}}^{\frac{h}{ct}} W_\theta(\theta) \cos(kr_r\theta)\, d\theta, \tag{68.5}$$

where $k = \omega_0/c$ is the wave number. Limiting ourselves to the case in which $W_\theta(\theta)$ is an even function, we rewrite (68.5) as

$$B_F(t, r_r) = 2P(t) \int_0^{\frac{h}{ct}} W_\theta(\theta) \cos(kr_r\theta)\, d\theta. \tag{68.6}$$

Inasmuch as $B_F(t,r_r)$ and $W_\theta(\theta)$ are related to each other, in accordance with (68.6), by a Fourier cosine transform, then, for the unknown function we have, by inversion

$$W_\theta(\theta) = \frac{k}{2\pi P(t)} \int_0^\infty B_F(t, r_r) \cos(kr_r\theta)\, dr_r, \tag{68.7}$$

in which case $W_\theta(\theta)$ is computed for the interval of values $\theta \in [0, h/ct]$.

Thus, the angular spatial distribution of scatterers, $W_\theta(\theta)$, is found from the spatial (vertical) correlation function of reverberation, using a Fourier transform.

CONCLUSION

The selected problems of applying probability and statistical methods in sonar presented in this book demonstrate the high effectiveness of the machinery behind the theories of probability and mathematical statistics in developing and analyzing models of signals and interference, in planning experiments, and in processing sonar information. In this book, most attention was focused on obtaining analytical results for the problems discussed.

In conclusion, we will discuss some problems of statistical theory of sonar that still await solution.

In developing probability models of signals and interference we maintained a phenomenological approach as a matter of course without examining wave models of sonar information. At the same time, combining the wave and phenomenological approaches may turn out to be extremely productive with respect to creating models which reflect real signals and interference adequately and provide a complete probability description. This approach is detailed in [13,34,35,39; also, in particular, 63], stimulating the development of a whole series of new problems in this field, such as:

(a) development of probability models of signals propagating in layered heterogeneous and statistically heterogeneous media; cf. [39];

(b) development of probability models for long-range reverberation, the properties of which are defined by both the characteristics of scattering irregularities and the characteristics of acoustic wave propagation (multiple-beam in the general case);

(c) creation of methods by which to determine multidimensional probability densities or characteristic functions of sonar fields, including non-Gaussian fields [52,70].

In planning experiments and statistical measurement in sonar, we encounter a number of problems connected with the synthesis of applicable information processing systems. Three approaches have been outlined to the synthesis of systems (see, for example, [8,13,18,26]): the classical statistical, the adaptive, and the nonparametric approach. In the classical approach, optimization can be achieved precisely, but a complete probability description of input effects (signals and interference) is required for this purpose. Moreover, models of input effects change in time and space, such that the system turns out to be nonoptimal within a broad class of conditions. In the adaptive approach the system must be optimized with current information. Such algorithms are rather complex, and it is not always possible to prove convergence. In the parametrical approach a nonoptimal system is created beforehand. Although it may be nonoptimal, in the first place it is impervious to changes in models of input effects, and in the second place it is extremely simple to use.*

Selection of the method for optimization (classical statistical, adaptive, or nonparametric) must not, of course, be governed by the tastes of the investigator, as frequently happens, unfortunately. A strict, fully rational approach is possible in this case, which reduces to solving the following systemic problem: The correspondence between the volume of formalized *a priori* information available, the desired effectiveness of the sonar system, and the technical and economic limitations, must be determined. In any case, inasmuch as success in synthesizing active sonar systems depends in many ways on the selection of an efficient optimization method, this problem requires the most persistent attention of investigators.

We note some of the general tasks connected with the synthesis of sonar information processing systems:

---

*However, this robustness is usually achieved at a non-negligible loss of sensitivity *vis-à-vis* the other, optimal approaches. (D.M.)

(a) development of constructive ways to overcome *a priori* difficulties connected with the incomplete probability description of input effects;

(b) creation of methods for multiparametric system optimization;

(c) development of the theory of hierarchical systems, the synthesis of which can be carried out in stages, using different optimization methods.

The next group of problems involves the development of methods for statistical measurement and for solving inverse probability problems. These two directions are interrelated, primarily because statistical measurements are used as the basis for solving inverse problems. Consequently, in many cases, the correctness of the solutions to inverse problems depends on the proper organization and conduct of the measurements.

We note the following problems which must be solved which are connected with the development of the theory and methods of statistical measurement:

(a) creation of a constructive classification of random processes and fields which are compatible with the methods of statistical measurement;

(b) development of the methods of multidimensional statistical measurement;

(c) solution of problems involving methodological support for statistical measurement, primarily that of defining standard random processes;

(d) development of theory and methods for synthesizing statistical measuring systems.

The most important task in solving inverse probability problems in sonar is studying their correctness conditions. In this regard we should mention the following lines of investigation:

(a) research on the effects of errors (interference and distortions) in the initial data on the uniqueness and stability of the solutions;

(b) development of methods for statistical control of inverse problems;

(c) determination of the possible effects the inadequacy of a probability model assumed for an active phenomenon may have on the correctness of inverse problem solutions.

Problems of building the relevant systems designed for conducting experiments on the properties of underwater objects (characteristics of scattering irregularities, water medium boundaries, etc.) are important to the development of statistical methods in sonar, as well. As a component of an experimental research system, the sonar system must be developed and used with consideration for the best solution to the problem posed, on the one hand, and with consideration for permissible expenditures on building and using the system, on the other.

In the last few years a considerable amount of attention has been devoted in the literature to the technical aspects of systems and the problems connected with such aspects (see, for example, [5,7,13,29]). Unfortunately, the technical aspects of systems have not as yet been formulated into an adequately clear scientific-technical discipline. Nevertheless, if we ignore the general methodology of the problem and concentrate our attention on the practical problems of developing the relevant systems and planning the experimental research, we can define the range of problems associated with sonar systems engineering.

We note several points of importance to these problems from a systems viewpoint:

(a) the sonar system is a subsystem in a system of a higher order;

(b) quantitative indices describing the success in solving the problems posed are defined for the sonar system;

(c) resources (available equipment, time expenditures, material expenditures, etc.), which are always limited, are assigned to the creation of a sonar system and to the planning of experimental research;

(d) a quantitative characteristic of the effectiveness of the sonar system is defined for each type of sonar system. This characteristic takes account of both the success in solving the problems posed and the resources expended on building the system;

(e) probability models of input effects (signals and interference) and the conditions of experimental research are assigned when sonar systems are analyzed and synthesized.

The points thus enumerated enable us strictly to formulate and solve two problems connected with the planning and conduct of experimental research, namely, those of analyzing and synthesizing sonar systems. In this case, we define analysis as determining the characteristics of the effectiveness of the sonar systems for particular probability models of underwater situations, given that the characteristics of the system are preset. We define the synthesis of an optimum sonar system as selecting that one of its possible versions which corresponds to the highest value of the index of effectiveness.

On the basis of the range of practical problems enumerated above we can formulate the subject of sonar systems engineering.

We interpret sonar systems engineering as a field of science dealing with the analysis and synthesis of sonar systems on the basis of adopted quality and effectiveness characteristics.

Naturally, planning of experimental research in sonar lies within the bounds of systems engineering tasks.

We note that systems engineering methods have been developing intensively in the last few years in radar.

In his foreword to reference [7] Academician A. L. Mints notes that radar systems engineering is an engineering discipline primarily requiring application of the procedures and apparatus of technical-economic analysis. It is also noted in this work that simplification of description is the most fundamental aspect of research on complex systems, and systems engineering is interpreted as the "science of simplification," in the sense that any sort of formalized description -- that is, mathematical models of input situations, models of a system and its quantitative indices -- involves simplification of the properties of the actual objects of research.

Systems engineering planning of experimental research includes the following basic stages (the situation is similar for radar [7]):

(a) determining the purposes of experimental research, their formalization, and selection of functions describing the effectiveness of the sonar system;

(b) determining the structural diagram for conducting experimental research and the structural diagram of the sonar system as a subsystem;

(c) development of algorithms for operation of the system and its separate components, as well as algorithms for controlling the system;

(d) determining the resources (technical resources, time and material expenditures, etc.) needed in constructing the sonar system;

(e) adoption of a method for optimizing the system, determining and developing a method for computing the indices of its effectiveness;

(f) selection of the best version of the system (in the sense of having the greatest index of effectiveness) -- that is, solution of the optimization problem.

It is clear from this list of stages in systems engineering planning of experimental research that this field has a totally independent scientific-technical content and great practical significance.

BIBLIOGRAPHY

1. Aizinov, M. M., *Selected Problems of Signal Theory and Circuit Theory,* Moscow, Svyaz', 1971.

2. Aleksandrov, I. A., and Ol'shevskii, V. V., A statistical description of some types of non-Gaussian processes in sonar, *Proceedings of the First All-Union School-Seminar on Sonar*, Novosibirsk, Nauka, 1970, pp. 55-73.

3. Blok, A. V., and Ol'shevskii, V. V., On the use of a probability model of a nonstationary random process for adaptive optimization of statistical measurement procedures, *Proceedings of the Fourth All-Union Symposium "Methods of Presentation and Procedures for Analyzing Random Processes and Fields"*, Section I, Leningrad, 1971, pp. 18-26.

4. Brekhovskikh, L. M., *Waves in Layered Media*, Moscow, Izd. AN SSSR, 1957.

5. Vasil'ev, B. V., *Predicting the Reliability and Effectiveness of Radioelectronic Devices*, Moscow, Sovetskoe Radio, 1970.

6. Woodward, P. M., *Probability and Information Theory, with Application to Radar,* London, Pergamon Press, 1953.

7. Kontorov, D. S., and Glubev-Novozhilov, Yu. S., *Introduction to Radar Systems Engineering,* Moscow, Sovetskoe Radio, 1971.

---

*References 1-36 appeared in the Russian edition. References 37-70 have been added by the Editor. (D.M.)

8. Kotyuk, A. F., and Ol'shevskii, V. V., Methodology problems of random processes and fields, *Proceedings of the First All-Union Symposium "Methods of Representation and Procedures for the Analysis of Random Processes and Fields"*, Vol. 1, Novosibirsk, 1968, pp. 1-16.

9. Kotyuk, A. F., Ol'shevskii, V. V., and Tsvetkov, E. I., *Methods and Procedures for Analyzing the Characteristics of Random Processes*, Moscow, Energiya, 1967.

10. Lavrent'ev, M. M., *On Some Incorrect Problems of Mathematical Physics*, Novosibirsk, Izd. Sibirskogo Otdeleniya AN SSSR, 1962.

11. Makhonin, G. M., and Ol'shevskii, V. V., Some ways of decreasing error in the recognition of random processes in statistical measurement, *Proceedings of the Second All-Union Symposium "Methods of Representation and Procedures for Analyzing Random Processes and Fields"*, Vol. 2, Novosibirsk, 1969, pp. 97-102.

12. Middleton, D., *Introduction to Statistical Communication Theory*, Vols. 1 and 2, New York, McGraw-Hill, 1960.

13. Middleton, D., Multidimensional detection and extraction of signals in random media, *Proc. IEEE*, Vol. 58, No. 5, 1970, pp. 100-110.

14. Ol'shevskii, V. V., *Statistical Properties of Marine Reverberation*, Moscow, Nauka, 1966. [English translation: New York, Plenum Press, 1967.]

15. Ol'shevskii, V. V., On the effects of scatterer movement on the statistical characteristics of marine reverberation, and solution of the inverse problem, *Trudy Akusticheskogo Inst. AN SSSR*, No. 2, 1967, pp. 206-216.

16. Ol'shevskii, V. V., Probability distributions of reverberation signals in the presence of low scatterer density and coherent scattering, *Trudy Akusticheskogo Inst. AN SSSR*, No. 3, 1967, pp. 214-231.

17. Ol'shevskii, V. V., On nonstationary features of marine reverberation described by a discrete model of sound scattering, *Sixth All-Union Conference on Acoustics, Abstracts of Reports*, Moscow, ASh4, 1968.

18. Ol'shevskii, V. V., Adaptive methods for optimizing the measurement of the characteristics of nonstationary random processes, *Proceedings of the Second All-Union Symposium "Methods of Representation and Procedures for Analyzing Random Processes and Fields"*, Vol. 2, Novosibirsk, 1969, pp. 78-91.

19. Ol'shevskii, V. V., The variance of the instantaneous frequency-time cross correlation function of an emitted signal and marine reverberation, *Akust. Zh.*, Vol. 15, No. 2, 1969, pp. 261-264.

20. Ol'shevskii, V. V., Problems of developing mathematical models of random processes and fields, *Proceedings of the Third All-Union Symposium "Methods of Representation and Procedures for Analyzing Random Processes and Fields"*, Section II, Leningrad, 1970, pp. 3-12.

21. Ol'shevskii, V. V., Mathematical models and statistical description of sonar signals, *Proceedings of the First All-Union School-Seminar on Sonar*, Novosibirsk, Nauka, 1970, pp. 3-33.

22. Ol'shevskii, V. V., *Introduction to the Statistical Theory of Active Sonar (A Training Manual)*, Izd. Taganrogskogo Radiotekhn. Inst., 1971.

23. Ol'shevskii, V. V., Probability models, experimental research, and statistical measurement, *Proceedings of the Fourth All-Union Symposium "Methods of Representation and Procedures for Analyzing Random Processes and Fields"*, Section II, Leningrad, 1971, pp. 3-21.

24. Ol'shevskii, V. V., and Tsvetkov, E. I., On the meaning of terms used in the theory of machine analysis of random processes, *Proceedings of the Third All-Union Symposium "Methods of Representation and Procedures for Analyzing Random Processes and Fields"*, Section I, Leningrad, 1970, pp. 3-7.

25. Pugachev, V. S., *The Theory of Random Functions and Its Application to Problems of Automatic Control*, Moscow, Fizmatgiz, 1960. [English translation: London, Pergamon Press, 1965.]

26. Root, W. L., Introduction to the theory of signal detection in noise, *Proc. IEEE*, Vol. 58, No. 5, 1970, pp. 8-22.

27. Bark, A. S., Bol'shev, L. N., Kuznetsov, P. I., and Cherenkov, A. P., *Rayleigh-Rice Distribution Table*, Moscow, Computer Center of the Academy of Sciences of the USSR, 1964.

28. Tyurin, A. M., Stashkevich, R. P., and Taranov, E. S., *Principles of Hydroacoustics*, Leningrad, Sudostroenie, 1966.

29. Fedorov, V. V., *Theory of the Optimum Experiment*, Moscow, Nauka, 1971.

30. Horton, J. W., *Fundamentals of Sonar*, U. S. Naval Inst., 1957.

31. Chernov, L. A., *Propagation of Waves in a Medium with Random Irregularities*, Moscow, Izd. AN SSSR, 1958.

32. Shirman, Ya. D., and Golikov, V. N., *Principles of the Theory of the Detection of Radar Signals and Measurement of Their Parameters*, Moscow, Sovetskoe Radio, 1963.

33. Faure, P., Theoretical models of reverberation noise, *J. Acoust. Soc. Am.*, Vol. 36, February 1964, pp. 259-266.

34. Kincaid, T. G., On optimum waveforms for correlation detection in the sonar environment: reverberation-limited conditions, *J. Acoust. Soc. Am.*, Vol. 44, September 1968, pp. 787-796.

35. Middleton, D., A statistical theory of reverberation and similar first-order scattered fields, Parts I and II, *IEEE Trans. Information Theory*, Vol. IT-13, July 1967, pp. 372-392, 393-414.

36. Van Trees, H. L., Optimum signal design and processing for reverberation-limited environments, *IEEE Trans. Military Electronics*, Vol. MTL-9, October 1965, pp. 212-229.

37. Stewart, J. L., and Westerfield, E. C., A theory of active sonar detection, *Proc. IRE*, Vol. 47, May 1959, pp. 877-881.

38. Van Vleck, J. H., and Middleton, D., A theoretical comparison of the visual, aural, and meter reception of pulsed signals in the presence of noise, *J. Appl. Phys.*, Vol. 17, 1946, pp. 940-971.

39. Middleton, D., A statistical theory of reverberation and similar first-order scattered fields, Part III: Waveforms and fields; Part IV: Statistical models, *IEEE Trans. Information Theory*, Vol. IT-18, 1972, pp. 35-67, 68-90; also *Correspondence, ibid.*, Vol. IT-21, 1975, pp. 97-99.

40, Horton, C. W., Sr., *Signal Processing of Underwater Acoustic Waves*, Superintendent of Documents, U. S. Government Printing Office, Washington, D. C., 1969 (see also the Bibliography therein).

41. Stephens, R. W. B., *Underwater Acoustics*, New York, Wiley, 1970.

42. Tolstoy, I., and Clay, C. S., *Ocean Acoustics*, New York, McGraw-Hill, 1966.

43. Parkins, B. E., Scattering from the time-varying surface of the ocean, *J. Acoust. Soc. Am.*, Vol. 42, 1967, pp. 1262-1267.

44. Fortuin, L., The sea surface as a random filter for underwater sound waves, *J. Acoust. Soc. Am.*, Vol. 52, 1972, pp. 302-315.

45. Middleton, D., *Topics in Communication Theory*, New York, McGraw-Hill, 1965.

46. Helstrom, C. W., *Statistical Theory of Signal Detection*, 2nd ed., London, Pergamon Press, 1968.

47. Van Trees, H. L., *Detection, Estimation, and Modulation Theory*, (Parts I, II, III), New York, Wiley, 1968, 1971, 1971.

48. Albers, V. M. (ed.), *Underwater Acoustics*, Vol. 2, New York, Plenum Press, 1967.

49. Morse, P. M., and Ingard, K. U., *Theoretical Acoustics*, New York, McGraw-Hill, 1968 (especially Chapters 8, 11).

50. Green, P. E. Jr., *Radar Astronomy Measurement Techniques*, Tech. Rpt. No. 282, M.I.T. Lincoln Laboratory, Lexington, Mass., December 1963.

51. Kennedy, R. S., *Fading Dispersive Communication Channels*, New York, Wiley, 1969.

52. Middleton, D., Probability models of received scattered and ambient fields, *IEEE, Engineering in the Ocean Environment*, International Symposium at Newport, Rhode Island, September 1972.

53. Plemons, T. D., Shooter, J. A., and Middleton, D., Underwater acoustic scattering from lake surfaces. I. Theory, experiment, and validation of the data, *J. Acoust. Soc. Am.*, Vol. 52, November 1972, pp. 1487-1502; II. Covariance functions and related statistics, *ibid.* pp. 1503-1515 (see, in particular, Ref. 26, Part I, therein).

54. Jahnke, E., and Emde, F., *Tables of Functions*, New York, Stechert, 1938, pp. 275-282.

55. Slater, L. J., Confluent hypergeometric functions, Section 13, pp. 503-535 in *Handbook of Mathematical Functions*, Abramowitz and Stegun (eds.), National Bureau of Standards (U. S. Department of Commerce), Applied Mathematics Series, No. 55, June 1964.

56. Moose, P. H., *On the Detection of Signals in Reverberation*, Doctoral Dissertation, University of Washington, Seattle, 1970.

57. Swarts, R. L., *Covariance Function of the Acoustic Back Scatter from a School of Fish*, Master's Thesis, Electrical Engineering Department, Oregon State University, Corvallis, 1971.

58. Ehrenberg, J. E., and Lytle, D. W., Acoustic techniques for estimating fish abundance, *IEEE Trans. Geoscience Electronics*, Vol. GE-10, 1972, pp. 138-145.

59. Horton, C. W. Sr., A review of reverberation, scattering, and echo structure, *J. Acoust. Soc. Am.*, Vol. 51, 1972, pp. 1049-1061.

60. Derr, V. E. (ed.), *Remote Sensing of the Troposphere*, U. S. Government Printing Office, Washington, D. C., (especially Chapter 29, "Channel modeling and system optimization for remote sensing").

61. Deuser, L. M., *An Environmentally Adaptive, Nonparametric Approach to Some Classification Problems in Underwater Acoustics*, Doctoral Thesis, December 1975, University of Texas at Austin (see *IEEE Trans. Information Theory*, Vol. IT-21, July 1976, pp. 501, 502 for an abstract).

62. Jobst, W. J., and Smits, T. I., Mathematical model for reverberation: experiment and simulation, *J. Acoust. Soc. Am.*, Vol. 55, 1974, pp. 227-236.

63. Middleton, D., A new approach to scattering problems in random media, International Symposium on Multivariate Analysis, Wayne State University, June 1975, Invited Paper; also, in *Multivariate Analysis* Vol. IV, P. R. Krishnaiah (ed.), North Holland Publishing Co., 1977, pp. 407-430. See, in addition, Ref. 1 therein and papers presented at the 89th Meeting of the Acoustical Society of America, April 1975 (papers BB5, 6); 90th Meeting, November 1975 (paper HH7). For related work, see the above and also a series of invited lectures (1973, 1976) at the Acoustics Institute, Academy of Sciences of the USSR, especially D. Middleton, Statistical-acoustical models of the ocean, and D. Middleton and V. V. Ol'shevskii, Contemporary problems in the construction of acoustical-statistical models of the ocean, together with a number of relevant papers by others, in *Trudy Akusti-*

cheskogo Inst. AN SSSR, late 1977. This work represents the development (since 1972) of a new theory of scattering for (linear) random media (to be published in *J. Acoust. Soc. Am.*, in 1979).

64. Tatarskii, V. I., *The Effects of the Turbulent Atmosphere on Wave Propagation*, Vol. T.T 68 - 50/464, Published for NOAA and NSF, Washington, D. C. (1971), NTIS (Department of Commerce), Springfield, Va.

65. Middleton, D., Doppler effects for randomly moving scatterers and platforms, *J. Acoust. Soc. Am.*, Vol. 61, No. 5, May 1977, pp. 1231-1250.

66. Grace, O. D., and Pitt, S. P., Quadrature sampling of high-frequency waveforms, *J. Acoust. Soc. Am.*, Vol. 44, 1968, pp. 1453-1454 (L), and references in [53].

67. Cook, C. E., and Bernfeld, M., *Radar Signals*, New York, Academic Press, 1967.

68. Bendat, J. S., and Piersol, A. G., *Measurement and Analysis of Random Data*, New York, Wiley, 1966.

69. Steinberg, B. D., *Principles of Aperture and Array Design*, New York, Wiley, 1976.

70. Middleton, D., Statistical-physical models of electromagnetic interference, *IEEE Trans. Electromagnetic Compatibility*, Vol. EMC-19, No. 3, August 1977, pp. 106-127.